BICYCLING

with

BUTTERFLIES

BICYCLING

with

BUTTERFLIES

MY 10,201-MILE JOURNEY FOLLOWING

THE MONARCH MIGRATION

SARA DYKMAN

PORTLAND, OREGON

Published in 2021 by Timber Press, Inc.
The Haseltine Building
133 S.W. Second Avenue, Suite 450
Portland, Oregon 97204-3527
timberpress.com

Text design by Vincent James
Jacket and map design by Faceout Studio
Printed in the USA
on paper containing 30% post-consumer waste.

ISBN 978-1-64326-045-7
A catalog record for this book is available
from the Library of Congress and the British Library.

To the monarchs

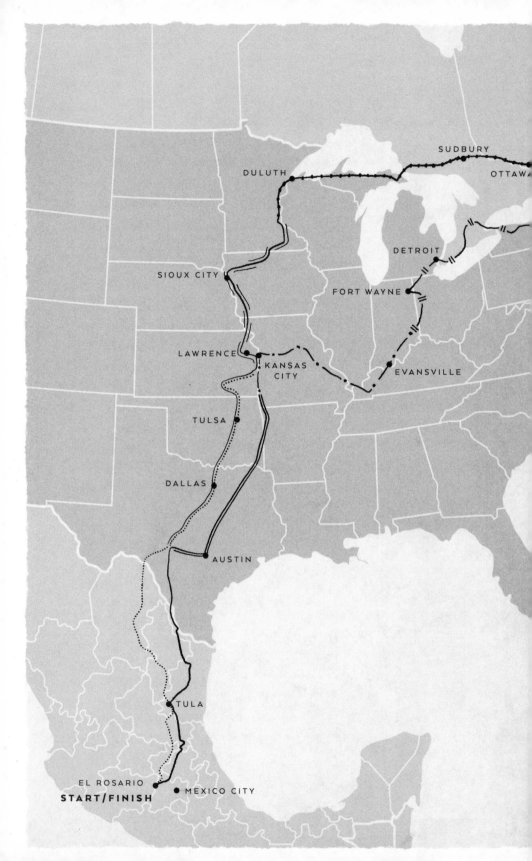

KEY

············· MARCH

·················· APRIL

MAY

JUNE

JULY

AUGUST

SEPTEMBER

OCTOBER

NOVEMBER

BOSTON

NEW YORK

THE EASTERN MONARCH MIGRATION

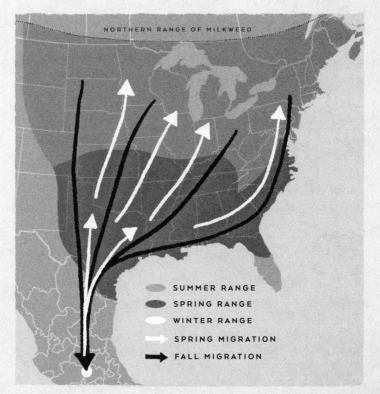

NORTHERN RANGE OF MILKWEED

SUMMER RANGE
SPRING RANGE
WINTER RANGE
SPRING MIGRATION
FALL MIGRATION

Contents

Preface

Scanning the forest's belly, I paused for a second look at some distant branches sagging with the weight of what looked like hives. *No, not hives*, I thought. *Are those nests?* Beyond my footsteps, mahogany-colored clumps dripped from the branches of scattered trees. They shimmered in the sparse light, nearly indistinguishable from the tree trunks and pine needles.

Wait, are those . . . ?

Puzzled, I ambled closer. Through squinting eyes, I began to see the details made clear by the light and shadow that grazed each wing and defined each butterfly. Not just any butterflies, and not just one. Millions. Millions of monarch butterflies. Known to scientists as *Danaus plexippus* and to Spanish speakers as *la mariposa monarca*; to me they were simply spectacular. I stood, enchanted. I had seen pictures and videos, but now I was finally among them. Millions—clinging to the trees like shelved books waiting to be read, their stories of adventure painted on their wings. Each had flown thousands of miles to escape the freezing winters of the United States and Canada. Each had the potential to travel many more miles back north in the spring. As did I.

Soon, as had happened for so many springs before, the monarchs would leave the protection of their canopy shelter and launch north. Unlike all the springs before, however, I would go with them: the first person to ever attempt to bicycle the entire route of the monarch butterfly migration. I stared up at my future traveling companions. They huddled in silent bundles on the branches and coated the tree trunks in stilled wings. Those trees, wearing butterfly wings, would function as the start and finish line of my upcoming adventure. When warmer weather nudged the monarchs to the sky, I would begin.

Arriving at the Start

The idea to bike from Mexico to Canada and back with the migrating monarch butterflies arose from a simple wish to visit them. In 2013, crossing Mexico by bike for the first time, a friend and I entertained the idea of visiting the monarchs at their overwintering sites. Because it was April and the monarchs had already begun migrating north, we decided to forego the side trip.

I spent the next few years idly daydreaming about returning. Over time, my plan morphed and grew—until I no longer wanted to just visit the migrants, but to accompany them by bicycle on their great migration. In 2016, I stopped daydreaming and picked a start date for my journey: spring of 2017. My idea was now a plan, and I had a year to work out all the details.

As with every adventure, planning was part of the fun. For a year I immersed myself in emails, web design, press releases, and business cards. I talked with scientists, clicked through websites, pored over maps, questioned my plan, and traced the vague outline of a route.

Eventually, there was nothing left to do but start. In January 2017, I braved a fifty-two-hour bus ride from my hometown outside Kansas City, Kansas, followed by a two-day bike ride, to arrive at the parking lot of the El Rosario monarch sanctuary in Michoacán, Mexico.

Including El Rosario, Mexico shelters between seven and eighteen known overwintering monarch colonies every winter. The number varies because smaller colonies are not consistently occupied and new colonies are still being discovered. Four of the colonies are open to the public: Piedra Herrada and Cerro Pelón in the State of Mexico, and Sierra Chincua and El Rosario in the neighboring state of Michoacán. Though the designations of "colony" and "sanctuary" are often used interchangeably, the listed public sites, aside from El Rosario, are technically the names of *sanctuaries* containing specific monarch *colonies*. El Rosario, on the other hand, is technically the name of a colony, found in the sanctuary Sierra Campanario. If such nuances of nomenclature confuse you, don't worry. I too was confused. Plotting my route prior to my arrival, I found just locating these sites, with their different names, difficult. Yet once in Mexico, there was always a local to help point me in the right direction.

Call them sanctuaries or colonies, I was able to visit the resting monarchs at all four public locations, and note their differences. El Rosario was the most built up, with multiple parking lots, many souvenir stands, and a paved trail to start (including 600 cement stairs). Along with Piedra Herrada, El Rosario also had the largest crowds of people, especially on the weekends. Cerro Pelón, on the other hand, felt the most like wilderness, with a long, steep slog to the colony. I found Cerro Pelón's trail to be the most difficult and Sierra Chincua's single track, with its mellow grades that dipped up and down, the easiest to negotiate. The subtle variations of each colony's forest composition yielded dozens more distinctions, from the ratio of pines to firs, to the openness of the flowering understory. Each colony was part of the whole, yet each was unique. Each *felt* different. Everyone who visits develops a favorite.

Arriving in Mexico in January, I chose El Rosario for my first visit not only because it consistently has the most monarchs, but because it is the most accessible. I arrived at the parking lot, walked under the arched entrance, bought an entrance ticket for fifty pesos (US $2.50), and met my guide, Brianda Cruz Gonzáles. Together, we began walking up the trail.

Had it been an option, I would have opted to go alone up the mountain. But one of the rules at the overwinter sites is that visitors must be accompanied by a local guide. Most days at El Rosario there were around seventy guides waiting to lead hikers up the mountain, and forty more waiting to take people up on horseback. Besides keeping a watchful eye on both tourists and monarchs, such work provides local economic opportunities and reduces the pressure on the mountains and forest to provide logging, mining, and cultivation jobs. The guides are a mix of young and old, men and women; it was my good chance to have been paired with Brianda. She was twenty-six and lived with her family at the outskirts of town, where there were more fields than houses.

I didn't know it on that first visit, but soon I would come to see Brianda as a sister and her house as my own. I would come to learn that among friends, she had a perennial smile, a hearty laugh, and a strength that, despite our different worlds, I saw in myself. As I walked with Brianda that first morning, in the company of towering oyamel firs (*Abies religiosa*) and leggy, smooth-barked Mexican pines, she was just my guide. She patiently led me down a dusty trail, forgave me for my crummy Spanish, and courteously laughed at my attempted jokes. "Respiro profundamente solo porque quiero," I explained. We both chuckled despite the fact that my joke, "I'm only breathing hard because I want to," wasn't that funny. I was simply acknowledging, with a bit of self-deprecation, two truths. One: I was out of shape and breathing hard. At 10,000 feet above sea level, my Midwestern lungs craved the missing oxygen. Two: I *wanted* to breathe hard. I wanted to feel my body striving upward through the forest. I liked that to seek out the monarchs, one had to struggle a bit. Beautiful sights are made more beautiful by the challenge of getting there.

As if on cue, a hummingbird rocketed through the understory to investigate the long, red flowers bent toward the hazy sun. The forest was bathed in salvias: both the large, trumpet-shaped, red stalks and the smaller purple flowers. I paused to catch my breath and turned my gaze upward. I was still unsure of what millions of monarchs clustered together really looked like. All I knew was that treasures were not easy to find, and that winter's beauty was guarded by vast space, steep mountains, and the echoes of a long-standing forest.

That forest, the last remnants of Mexico's expansive, ancient ecosystem, caps twelve isolated massifs clustered in a volcanic mountain chain in central Mexico. After the last ice age, as temperatures increased, the oyamel fir forest that had once covered much of southern Mexico was forced to retreat to the cool, humid refuge of the mountains' higher elevations. The oyamel firs of today are the witnesses of yesterday. Looking up at their branches—a net to catch the sky—I understood why they were also known as the sacred fir. Their branches *did* seem to splay out like many crosses, and their soft, short needles looked like fingers folded together in prayer. A church of trees, contracting with time. They were a reminder that as the planet changes, nothing is forever.

Once sprawling, the high-elevation oyamel holdouts now occupy less than 0.5 percent (approximately 100,000 to 124,000 acres) of Mexico. In comparison, the 2010 United States Census put Kansas City at 201,568 acres. Despite the limited area, the monarchs arrive each winter, and the oyamel fir forest absorbs nearly every monarch born between the Rocky Mountains and the Atlantic Ocean. It is a concentration of monarchs that saturates the trees and transforms the forest into the focal point of the range, an orange gem strung on a volcanic necklace.

An hour after we started hiking, Brianda signaled toward hive-like nests dangling in the branches. I stood puzzled. Then, like a stereogram image, the bundles began to define themselves. The monarchs came into focus. Their collective weight bent each branch into an archway. I stepped forward, but did not enter. Instead, I craned my neck upward to contemplate each tree shrouded in monarchs, while the monarchs, like monks, contemplated winter.

Their weight fell onto my eyes until I shut them. I had arrived at the start of my trip, the start of my dream: to follow the monarchs by bicycle and give voice to their alarming decline. Now I had six weeks to wait for spring to bloom and the cold grasp of winter to loosen. Even in Mexico, it had a hold.

Despite Mexico's reputation for deserts and heat, in the high-elevation forest lit by a cloud-tangled sun, freezing storms and cold temperatures still bully the monarchs each winter. It is thanks to the protective scaffolding of the forest that the monarchs find literal and figurative sanctuary. The canopy, with its weave of branches, moderates temperatures (like a blanket) and shields precipitation (like an umbrella). The tree trunks absorb even mild heat each day, acting like warm water bottles that the monarchs can snuggle against in moments of extreme cold. At night, trunks tend to be an average of two and a half degrees Fahrenheit (F) warmer than the surrounding ambient night temperature.

These butterflies occupy a sliver of habitat speckled with microhabitats, seemingly scripted for their survival.

It is a balance steadied by Earth's many layers, and a balance tipped by humanity.

Each time a tree falls in the monarchs' overwintering forest, a hole is torn in their blanket and punched through their umbrella. These disturbances—logging, disease, windstorms, fire—allow heat to escape and moisture to enter, creating a dangerous combination.

Careful with my steps, I knelt to watch a winter-battling monarch crawl toward me. I knew he was a male because of his thin, black veins and the two small, black dots (scent glands) on his hind wings. I knew he was cold, because as he crawled, he shivered.

Monarchs are ectothermic (cold-blooded) animals. Their body temperature matches that of their environment. The colder the temperature, the colder monarchs get, and the more inactive they become. For much of the winter, being cold is an energetic advantage, yet, if monarchs get too cold they risk freezing. They must employ strategies to limit exposure to the coldest extremes. For this reason, monarchs tend to occupy the sunnier, south-facing slopes of the forest, and they form clusters under the forest canopy. In such clusters, they are protected by both the trees and the butterfly bodies that make up the outer edges of each mass (a bit like penguins).

The challenges of the cold are most acute for ground-stranded monarchs. As temperatures drop, ectothermic monarchs become unable to move and can't seek out microclimates, such as tree trunks. Monarchs must be at least 41 degrees F to crawl and 55 degrees F to fly (known as their flight threshold).

The monarch at my feet was just warm enough to crawl; he was shivering to warm his muscles to make an escape possible. Though slow, if he could climb even one foot off the ground, he could greatly increase his chances of survival. The ground held the coldest microclimates and the possibility of dew, plus the ever-present danger of black-eared mice (*Peromyscus melanotis*). It was a risky place for a monarch to pass the night.

Avoiding the coldest, wettest conditions is of the utmost importance for monarchs, as their nightmare scenario occurs when the two conditions overlap. Cold, dry monarchs at least stand a chance. Cold, wet monarchs are in real danger. Monarchs get wet when they are exposed to precipitation or dew. Clustering monarchs in a healthy forest are protected from storms, but as trees are removed, monarchs are left exposed. Monarchs on the outer edge of the clusters, especially those on the windward sides, can become coated in moisture. Monarchs that are blown off the clusters or those already on the ground when temperatures drop are even more at risk. Not only are they exposed to the coldest microclimates, but they can become coated in dew that forms most nights. Such moisture can then freeze on the surface of wet monarchs, creating ice crystals. As these ice crystals radiate out, they can spread not just across but inside the monarch, through the holes (called spiracles) they use to breathe. Freezing from the outside in is the deadly process known as inoculative freezing. In a field lab, wet monarchs exposed to declining temperatures all died by the time the temperature had fallen to 18 degrees F. At the same temperature, only 50 percent of the dry monarchs died.

The monarch at my feet shivered, but at least he was moving. I cheered him on, wishing I could offer him a cup of hot tea or a jacket. Instead, I settled on guarding him from oblivious tourists. In pantomime, I caught the attention of a group looking up, their footsteps unguarded, and reminded them to tread slower, more deliberately. Brianda, in the meantime, had found a stick, which she offered to the monarch as one might offer a hand to a dance partner. The monarch accepted. He gripped the stick, still shivering, and Brianda moved him off the trail.

For all the danger the cold entails, it is also a saving grace. Low temperatures keep the monarchs inactive. Instead of flying around and burning lots of calories, when cold, they can dangle from the trees, use very little energy, and conserve their fat reserves for their remigration north in the spring. Like nearly frozen statues, monarchs wait out winter in a hibernation-like slumber.

As an endothermic (warm-blooded) human, my temperature needed to remain stable despite the cold outside temperatures. Watching the millions of monarchs sleep, the cold nestled against my skin and I shivered. Shivering, like diverting blood from extremities and increasing metabolism, helps endothermic animals maintain a warmer temperature in the cold. My body confirmed the science. I zipped up my jacket, amazed that the monarchs had found this perfectly chilled forest.

I was not alone with the cold and the monarchs. Around me, other visitors huddled together. Since disturbances could send the monarchs into flight, and use their precious energy, there were a few rules: no touching the butterflies, no flash photography, and no talking. The nearly wordless crowd gave the forest the air of a church instead of a zoo. The forest felt like a temple made by trees and worshipped by a congregation of wings folded in prayer. What they prayed for I could only guess: tailwinds, milkweed, or the peace that exists in quietness. I joined them, praying in my own way for the strength to be part of the migration and battle the many miles ahead.

How many miles, I wasn't exactly sure. I figured I would need to bicycle around 10,000 miles if I wanted to go from the overwintering grounds in Mexico to Canada and back. If I left in March, I could get to Canada by summertime and be back in Mexico by November, just like the monarchs. That translated to a very plausible 1200 miles a month.

In the hushed forest with hushed monarchs, I tried to hush my doubt. I reminded myself that ever since I had been biking, even as a child making laps around the block, I had been proving to myself that I could go the distance. Laps around the block, then the neighborhood, then the city, trained me for my first bicycle tour when I was seventeen. That tour, a month of forty-mile days up the East Coast, taught me a useful truth: a long trip is nothing more than a collection of miles. If I could bike one mile, then I could bike two. If I could bike two, then I could bike 10,000.

Knowing I could cover the miles was not the same as thinking they would be easy. I knew that I would struggle, that some miles would hurt, that there would be moments of many extremes, and that the scope of it all could overwhelm me. It was not the 10,000 miles that I braced myself for, but the doubt that lurked in the vague outline of those distant miles.

For the monarchs, their very survival was in doubt. Yet, as their population staggered toward extinction, and uncertainty prodded my brain, the monarchs above me seemed peaceful, unburdened. They had been proving themselves, year after year, for thousands of years. I assumed they didn't appreciate this, nor could they comprehend the significance of their uncertain future. Yet it brought me comfort to imagine that they hung peacefully because they understood the bigger picture: that their job was to migrate across a continent—battle storms, predators, disease, human development, busy roads, and pesticides—until it wasn't. I took a deep breath, trying to put the present into the context of history. Surely if a butterfly with nothing more than instinct and orange wings could navigate three countries and the chaos of humanity, then I, with my stubborn will and a continent's worth of hospitality, could too.

It was only January. The future would come, as would each mile. In the meantime, I turned my attention back to the branches. There was little room to worry in a forest painted with monarchs.

The Monarchs' Winter Neighbors

On a cool morning in March, a few days before I began heading north, I paused to take in my surroundings. I had been in Mexico for nearly two months, and although my journey had yet to commence, the adventure was already underway. Squinting from the sun, I gazed out.

Above me was a monoculture pine forest planted for fuel and lumber. Such farm-like forests were common in this area, on the Michoacán–Mexico state border. Though not exactly like the native forest, the rows of pines were a profitable crop that prevented landslides and took the pressure off the forest higher up the mountain. Below me, on a spread of undulating hills, lay a distorted checkerboard of tilled fields still used for subsistence planting. Cement houses sat pastured. Lines of clothes dried in the sun and painted each house a dozen swaying colors. The main road extended like a dusty river downward. Its tributaries of footpaths reached every corner. Brianda's house, where I was staying, was close enough that I could see the family dog, Dobber, lounging in the sun. The neighbor's entitled turkeys cast clucking shadows en route to stealing seedlings from the family garden.

After my first visit to El Rosario, I had asked Brianda if she knew of any volunteer opportunities. I knew then that I had several weeks until the monarchs began flying north, and I wanted to help in whatever way was useful. Trail maintenance, monarch education, English lessons? She eagerly accepted my offer to teach English. Not only that, I was invited to stay at her house.

Now I was in a field that clung like a postage stamp to the steep hillside above her house. Like methodical ants wearing team jerseys made from dust, her family and I paced their field. With each step, we slowly filled the fresh furrows with next year's beans and corn. In just a few weeks, the family alongside me had become an extension of my own. Brianda had become my *hermanita*.

I let the beauty of the moment follow me as we continued our work. The sweaty horses, driven by Brianda's dad, Israel, quivered from their burden. Brianda's mom, Leticia, followed the fresh scar of the plow's wake, leaving three corn kernels in each of her wide-spaced footprints. I clumsily placed one of the multicolored beans between each pile of corn. Brianda's sister, Diana, covered the seeds with soil; her brother, Ivan, added some fertilizer; and their cousin tucked the seeds in with a final blanket of dirt. Our work lay hidden, waiting for the rains to come.

The monarchs were like a calendar. In April, after the monarchs left for the north, the beans and corn would begin to grow, taking advantage of the remaining humidity trapped in the soil. In late May, the rains would start, and the well-established crops could embrace the deluge. In November, when the monarchs returned, the crops would be ready to harvest. By reading the monarchs, one could read the rain and the growing cycle.

It was the intimacy of that moment that struck me most, of being part of the less glamorous rituals of life. I took the opportunity to soak up the gift they had given me: the gift of ordinariness, a window through which to see their unguarded lives.

As the waves of dirt spilled into the grassy edges of the field and the dust camouflaged our bodies, we continued our labor. In strict concentration, I raced to keep up and accurately place the beans. Every few minutes I would reflexively say "oops," as a rogue bean slipped from my hands and planted itself by disappearing into the soil.

"Ustedes saben que significa 'oops'?" I asked, wondering if oops meant the same thing in Spanish as it did in English.

"Si," they laughed, confirming that they knew exactly how bad I was at planting beans. Their laughter was acceptance. My smile translated my feelings of good fortune.

From tourist to blundering bean farmer, my journey so far had been one of luck—the kind of luck I aimed for by having only the vaguest of plans and a bare-bones schedule. I preferred the details—such as where I would sleep or buy food or shower—to be revealed only as needed (and conveniently, I could go days without *needing* a shower). There was a joyful freedom in not having much of a plan. I was free to eat when I was hungry, rest when I was tired, and camp when I deemed each day done. The details were unnecessary. For me, it was enough to know that I would end up somewhere.

This wing-it philosophy has been both a blessing and a curse throughout my travels. There have been days when not planning meant eating nothing but crackers and ketchup, biking miles of road unfit for cyclists, and camping in make-do sites. These challenging moments were likely avoidable, but the truth was, I did not want to avoid them. I wanted to invite them, and by not overplanning, I did just that. I tried to give space to that which I couldn't yet imagine, to encourage just enough discomfort for an adventure to unfold.

Perhaps the biggest advantage of my figure-it-out-as-I-go strategy was being able to say yes when opportunities presented themselves. It was because I had time to spare and only a vague plan that I was able to accept Brianda's invitation to stay with her family—an invitation that allowed me to not only plant a field with beans, but to become Brianda's shadow and learn what it was like to be a guide at El Rosario.

Six days a week during my stay, Brianda would rush out the door to greet the frigid dawn as well as her neighbor and coworker, Priscilla. Down the dusty dirt road we headed. They would fill the distance with jokes I didn't get and stories told faster and faster in Spanish as the plots thickened. Not a morning person, I let the Spanish float by untranslated. We would walk until transportation arrived. Five-passenger cars fit ten, and as the car's belly scraped at each speed bump, we would joke about losing weight. Trucks fit many more, and no one was too old or too young to hop in or out of the bed.

Arriving at El Rosario before the crowds, most of the guides would take a knee before the small shrine at the parking lot's far corner, then veer uphill to the main entrance. After going from guide to guide with a handshake and a "Buenos dias," I would set up shop in the cafeteria. There, as we huddled over cups of steaming *café de olla*, *atole*, or instant coffee, I would chat with the guides before pulling out the poster board of English phrases and launching into practice. Mostly, the guides rotated in and out, and I would be there to catch them in a moment of bravery to help them learn a new phrase. "My name is . . ." "Thank you." "Do you want to ride a horse?"

Most afternoons I would charge up the hill, half for exercise, half to say hi to the monarchs. Each year the butterflies are in a slightly different clump of forest. For most of the 2016–17 winter they were past the undulating forest trail that spilled into the Meadow of the Rabbits, El Llano de los Conejos. On my first visit we veered right, about ten minutes from the meadow. On subsequent visits, they had moved down a trail twenty minutes beyond El Llano. I could cover the four-hour (for most visitors) trip in forty minutes when the clouds were out and I didn't have to worry about trampling monarchs. On sunny days, I would linger to watch the delirious melding of orange wings and blue sky. On weekends, when the trails were overrun by thousands of weekend warriors, I would meander on side trails, skid trails, horse trails, cow trails, and through trailless meadows. On the days it rained or hailed, I would wait anxiously with the guides at the entrance, willing the monarchs to be safe and the storms merciful. I hoped 2002 would not be repeated.

January 11–16, 2002, a severe winter storm struck the monarch colonies. The first blow was forty-eight hours of rain. The second blow was a prolonged period of freezing temperatures. The deadly wet-cold combination was realized. An estimated 200 to 275 million monarchs, or 75 percent of the population at El Rosario and Sierra Chincua, were killed. On the ground, dead butterflies formed a mass grave. At some places the monarchs measured thirteen inches thick, and survivors, insulated by the dead, were unable to crawl out. The historic storm had been brewing for many years. The climate was changing, and trees that had insulated the forest and protected the monarchs from precipitation had been cut down. When the cold front came, the monarchs were exposed, and all but the most protected in the densest clusters fell victim.

That year, Estela Romero, a local woman from Angangueo who had been captivated by the monarchs since she was a little girl, traveled to several of the sanctuaries and collected hundreds of dead monarchs in the storm's aftermath. She placed the monarchs in a clay urn and took it to the community cemetery. On a visit to Angangueo, I wound my way through the cemetery's labyrinth of cement headstones and flower tributes to the far back corner, where a wall had begun to crumble. With respectful steps I located the marker I had come to see. Kneeling to brush away the pine needle duff, I read aloud, first in Spanish, then in English, then in French, the engraved tribute. "In remembrance of the millions of monarchs that froze to death in the great storm of January 2002."

The words settled around me, and when silence returned, I was alone with the memory of millions. I bawled. It was the first time I had gone to a cemetery and felt more than awkwardness. I brushed away more pine needles, desperate to convey with action my sorrow for such loss. It was cathartic to have a place to mourn for the planet. As the world continues to erode, we need places to say we're sorry. Places to remember what we have lost.

It wasn't just 2002 that saw catastrophic mortality at the overwintering colonies. In recent years, moisture-heavy storms have wreaked even more havoc. Historically, storms at such scale were unknown, but climate change has invited destruction. The winters of 2004, 2010, and 2016 all brought unprecedented and deadly storms, influenced by climate change and the warming of our oceans.

Oceans are powerful affecters. They create climate, act as thermal reservoirs to stabilize day and night temperatures, and redistribute heat by way of currents for a hospitable balance between the equator and the poles. As greenhouse gases build in our atmosphere, they trap more and more of the sun's heat. The planet warms, and the oceans absorb this warmth. As the Pacific Ocean heats up, evaporation increases, causing brewing storms to carry more and more moisture. Monarchs, which have evolved to survive the *dry* season in Mexico, are finding the dry season no longer exists. The monarchs are now drenched every few years in catastrophic winter storms. Adding insult to injury, the forest canopy is now less intact and can no longer offer reliable shelter.

One of the best ways to mitigate the damage caused by these new threats is to have a robust population. The monarchs can deal with major fluctuations in their population, because they have a multigenerational summer breeding season, during which many eggs are laid. As long as there are survivors and the progeny are met with favorable conditions in their summer range, monarch populations can rebound. In other words, if a winter storm kills 200 million monarchs, but the population is a billion strong, then there is room to recover. If, however, the same storm hits and the population is only 200 million to start, the fatal result is obvious. Each time hail fell, while I waited at El Rosario with the guides, I did this simple math. When the storms passed, I would walk back up the hill to inspect the damage. Only when I could still see the ground would I sigh with relief.

Up and down I went, seeing the monarchs daily and never wandering more than five miles from the cafeteria at El Rosario. This was how I liked it. Because I was a familiar sight, I was allowed to discover these people of Mexico, whose lives were uniquely linked to the monarchs. There was Pablo, my bravest student, who practiced English without fear of messing up. By the time he blurted, "I am bat shit," we were good friends. I had no idea what he actually meant to say, but I knew it was safe for my reply to be a laugh. His response was to try again. He was in second grade when his dad died. His family had moved, and his formal schooling had ended then. Watching how he progressed with such limited resources (and a teacher who didn't know what she was doing), I could only imagine how far an education like mine could have taken him.

There was also Eric, the young, hip teenager with perfect English pronunciation. There were Sylvia and Cati, sisters I often confused (Cati made delicious bread and practiced saying "My name is" with devotion). Fabi kept a notebook with her collection of English phrases, and one day sat down to read me the Bible (in Spanish). Jorge was a guide who led visitors on horseback. He was studying to be a veterinarian, and together we attempted to translate an English textbook, only to discover that if I didn't know what cryoprecipitate was in English, I definitely didn't know its Spanish equivalent. By becoming part of the daily lives of the guides, I was able to peek into the world as they saw it. See the monarchs as they saw them. See the reserve as they saw it.

In 1986, the Mexican government created what is now known as the Monarch Butterfly Biosphere Reserve (MBBR), to protect the monarchs' overwintering forest. Unlike national parks in the United States, the MBBR is a collection of mostly communal lands. The two major systems that predominate the reserve are common across Mexico. They include indigenous *comunidades* (38.4 percent of the reserve) and *ejidos* (48.2 percent). While these lands are in essence federally owned, people live on them, and are entitled to work the land to generate income. Such rights were granted after the land redistribution efforts that followed the Mexican Revolution (1910–1920).

Understanding Mexico's land history helps put into context what happened when the communal lands were declared protected lands by presidential decree in 1986. Without any say from the ejidos and comunidades, it simply became illegal for them to log their forest parcels. As with many actions taken in the name of conservation and environmental protection, the least affluent were the most disrupted. In response, there was a great backlash. People feared their land was being taken away, so they logged the trees while they had a chance.

In answer to this retaliation (and a 1988 law that required community input when declaring protected lands), the government tried again in 2000. This time, with community participation, the reserve was enlarged and restructured. The isolated monarch hotspots first designated as protected were connected, expanding the reserve to its current size of 139,019 acres. To compensate for the loss of timber revenue (and incentivize forest conservation), the Monarch Fund was established. Each year the fund—a collection of government and private money from Mexico—is divided among the ejidos and comunidades with landholdings in the most protected and regulated parts of the reserve.

For the ejidos, the money generated by the Monarch Fund and tourism is currently distributed at the discretion of *ejidatarios*—community members and the main stakeholders in the ejido system. El Rosario is an example of an ejido in the MBBR, overseen by 261 ejidatarios.

At El Rosario, I would often pass by the cafeteria as the ejidatarios held their bimonthly meetings to vote on decisions and distribute tourism revenue. Once I was pulled in for introductions, and people began to joke that I needed to become an ejidatario too. I adamantly dismissed even a joking offer, because I knew how coveted such a title was. Even in big families, only one person—typically the eldest son—inherits the title.

Each ejidatario at El Rosario gets the chance to either work as a guide, pass the right to work to a family member, or sell the right. Even then, job security is minimal. Work is only seasonal, and because there are limited guide jobs, the group is divided into three rotating sub-groups, each eligible to work once every three years. Any time a tourist asked me what an appropriate tip was, I tried to explain this complicated system. A big tip was easier to stretch out over the years between work. Since the title of ejidatario is usually passed to a single family member, a lot of siblings and their families are left out of the opportunity.

For such communal land systems to be successful, there must be good communication among the members, as well as sufficient incentives and resources to enforce the collective good of both the people and the forest on which they depend. Easier said than done. Until the benefits of conservation outweigh the benefits of extracting resources, the forest will continue to take the brunt of human need.

Between English lessons, I began to see the vague outline of Mexico's unresolved land and forest issues. I saw that not everyone benefited from tourism or conservation funds, and that the forest was still threatened. I wandered the woods and saw swaths of stumps, like headstones, basking in sunbaked soils. I found a web of roads that broke through the understory and left every tree vulnerable. I heard stories, and was warned to never visit the forest at night, when protected trees were robbed under the mask of darkness. One day, in a car with a group of tourists, I listened as the driver told a story of putting his head where it didn't belong. He told the story to all the passengers, but through the rearview mirror, I felt he looked only at me.

While mostly in the shadows, these issues were given a spotlight in 2020, when media reported the deaths of two monarch conservationists. At the time, I was staying near El Rosario. It was a time of great sadness and unease. It was also, for me, a frustrating time. I watched the news—out of context and in some cases completely misreported—reach the United States and Canada and stir up a fear frenzy. I responded to people's worries more or less always in the same way: "I am safe. Rumors are circulating, but it is clear to me that things are more complicated than the news suggests."

The best I can say for now is that I do not understand the politics of the monarchs and Mexico, but I understand that it has a complicated dark side. Perhaps less dark than that of the chemical companies hiding the hazards of Roundup, pumping money into limiting protections, and stealing subsidies to erase the American prairie. Perhaps not. Each time I felt the dark side in Mexico, I stepped back, not ready to commit to the danger. I wanted to ride my bike with the monarchs, and add to the voices that could fight the shadows.

Regardless of the perils, perceived or otherwise, conservation efforts are effecting change. Daily patrols of the forest offer the trees vigilance, and large-scale illegal logging is at an all-time low. Small saplings, recently planted, grow hope for the future. Roads that once carried large trucks loaded with timber are closed. Researchers monitored trees and monarchs alike. Spurred into action, conservation efforts continue.

One beneficial effect of protecting the monarchs, at least for those living near the overwintering grounds, is to foster tourism. During the 2012–13 season, 72,591 people visited the monarchs in Mexico. These tourists bought not only their entrance tickets, but food, lodging, transportation, and guide services. Each transaction became a reason to protect the forest. When the community provides for the monarchs, the monarchs can provide for the community.

By protecting the forest, the people can also protect themselves. Logged hillsides, without roots to stabilize the soil, wash away in deadly landslides. Disturbed forests are more susceptible to fires, disease, and climate change. So, while logging brings short-term wealth, it also brings dangerous side effects. Preserving the forest offers long-term benefits, including tourism.

If well-managed, tourism could bring more sustainable wealth. The question, however, is how do you manage the thousands of visitors that visit the monarchs each winter? At least for now, tourism brings its own burdens, and the forest is forced to contend. Trails widen with the trampling of thousands of visitors and horses, creating corridors of erosion in the rainy season. Parking lots are paved and then become speckled with trash. A wake of noise and disturbance follows the masses. Excited tourists carelessly tread the ground, smashing monarchs and plants alike. How do we choose when and what to protect, sacrifice, or exploit? How do we choose, when the lives of both monarchs and humans are at stake?

I didn't have the answer when I was there, and I still don't. I was simply there to learn and try to make myself useful. I taught some English, helped translate when needed, washed dishes in the cafeteria. When persuaded, I wore a spare guide vest and led my own groups up the mountains. Once, on a three-hour trip, I earned a tip of 300 pesos ($15)—enough to treat a handful of guides to lunch and buy dinner ingredients for the house.

Once, on a four-hour trip, my tip was ten pesos (50 cents)—barely enough to cover the cost of the car ride back that evening.

Tip or no tip, each night we caught a car to the end of the paved road, then walked the remaining mile of dirt. Unlike in the mornings, trudging back up the hill in the evenings I was awake enough to participate in the life of the road. Few people were absorbed in phones, even fewer were in cars. Instead, the road was an extension of each home, a long living room where friends and family caught up. We would wish a good evening to everyone: moms with kids in tow, young boys leading horses piled high with firewood, and all the faces in the small shops that accompanied each house. Such shops were a wonderful convenience. If we needed a staple (or candy!), we would bellow a hearty "Good evening!" and wait for the person who lived at the house to arrive and sell what they had. Usually choices were minimal but sufficient. You could always count on there being candy.

Dobber was usually the first to greet us. Tail wagging, he waited as we climbed the stairs cut into the hillside below Brianda's house. Most of the time, the sun was but a dim reminder of the day and an oppressive cold would have begun to take charge. Brianda's mom, Leticia, was always at the house to greet us, and the warm drinks she had standing by were always revered. I couldn't get over my luck when after a full day of exploring the monarchs with my new guide friends, I would return to a home on a hill in Mexico and be served heaping plates of Leticia's finest cooking: *mole*, beans, chicken soup, *chile rellenos*, eggs, or chard and zucchini from the garden. By the time we ate dinner and the customary second dinner of tea and bread, there was only time for some conversation around the table, then off to bed. Like the monarchs, we would tuck ourselves into the warmth of many blankets, defending as best we could against the cold.

Between trips to El Rosario, I whooshed down the mountain on my bicycle, past Ocampo, to the city of Zitacuaro. Hurtling down the 4000-foot descent, I could cover the twenty winding miles into the heat and smog with ease. Leaving the main highway, Zitacuaro engulfed me. On streets choked with vendors, I passed markets extending through hidden doors. Across the city's main garden square, children with balloons chased pigeons. I followed the grid to cross town and emerged on the city's less-developed edge, a section that unfolded with a glorious downhill run. I knew each break in the speed bumps, the cycle of the stoplight, the problematic potholes, and which store would still have bread if I arrived late. Beyond the gate of the last turn, there was nothing left to do but coast down the dirt trail and sink into the welcoming silence of Papalotzin, a monarch education center founded by Moises Acosta.

Moises named his butterfly center Papalotzin after the indigenous Nahuatl people's word for monarch butterfly. The Nahuatl understood the monarch to be the only creature capable of traveling silently enough to bring the peoples' wishes of happiness to their goddess of joy and flowers. I had sensed the monarchs' sacredness and whispered wishes to them even before I knew such a legend. How could one not feel reverence, watching a monarch fly high and disappear into the pellucid sky? I hoped the butterflies could fly long enough to not only bring our wishes for their protection to the gods, but to every North American on whom their survival depended.

I spent much time at Papalotzin. In the silence, I could hear the songs of pygmy owls and the wings of passing egrets caressing the sky. During the day, Moises would arrive, and together we would water the plants, lead butterfly tours, and tend to the captive monarchs living in the butterfly house. Those monarchs, and all the monarchs living around Zitacuaro, were not migratory like the ones living higher up in the mountains. They were a resident population and could be teachers and ambassadors all year round.

Moises was a teacher as well. He taught me how to navigate the physical and cultural worlds of monarchs in Mexico, brought me to the local radio station to share my story, and shared his insights as a local naturalist. Still, I made sure to always use the formal Spanish conjugations when talking with him and let him explain his world to me on his own terms. He was a stoic man, and a bit of a mystery.

Whatever our relationship, I respected Moises a great deal. He was attempting to support himself by supporting the monarchs. He saw the monarchs not only for their potential to contribute to the local economy, but for their intrinsic beauty. His dad had taken him to see the monarchs when he was a boy and he still held that simple, youthful wonder in his heart. His passion helped people see and understand the monarch. With understanding comes appreciation. Appreciation motivates action.

It wasn't easy for Moises to love monarchs. Ten years before my initial visit, Papalotzin was burned down, after Moises had reported illegal logging. He had also been robbed several times as he patrolled the forest. Most think courage is stepping into the unknown, but for me it is continuing even when you know what is out there.

I admired Moises's bravery and commitment and felt so fortunate to have crossed his path. When I said goodbye at the end of each visit, a pang of sadness would touch me. Then I would instead focus on where I was going. I had two homes in Mexico. Brianda's house would beckon once again, and I would pedal up what I had coasted down.

I would also focus on the monarchs. An increase in their activity had begun in mid-February, as sunnier days had begun to lure thirsty butterflies out to search for water on the drying hillsides. Monarchs flowed downhill each morning and returned, thirst quenched, every afternoon. This streaming behavior was a signal that the migration was approaching. Like stir-crazy adventurers, the restless migrants met me farther and farther down the road. Every encounter, especially those a few miles from the main clusters, were reminders to me that the start was near.

My bicycle adventure was about to begin.

A Million-Winged Sendoff

MILES 1–118

The sun's warmth began to pour steadily through the branches, and the monarchs responded by opening their wings, every scale twinkling with gratitude. In the spotlight of spring, butterflies by the thousands sailed toward the sky as fluttering eruptions of orange. The monarchs crowded the skies, painted poems against the blue, and danced with the wind. The song of millions of wings hummed through the trees' needles, and I felt the anticipation in their swirling flight.

As they sought water, monarchs leapt, like a parting river, from the moist ground where they pooled along the trail. When a cloud passed across the sun, the shade and temperature drop signaled to the butterflies that they might be trapped away from their colony. Suspended in air, they would wait. If the clouds stayed, they could glide back to the protection of the clusters. If the sun re-emerged, the monarchs could return to their errands: swirling in search of mates, sipping nectar, absorbing the renewed warmth, or settling back to the streams for more water. Pooling, streaming, mating, and flying were all clues that the migration was gearing up. Winter was over.

It was time to pack my bags and ride into battle—the battle to save a great migration.

I loaded down my beater bike, a 1989 Specialized Hardrock, until it was so heavy I could barely lift it off the ground. A Frankenstein bike that

I had made five years earlier from a collection of used parts, it looked like a cross between a salvage yard and a garage sale. Its white and pink paint job was speckled with rust-colored dings—scars from past adventures. The bike was ugly. To me, however, it was a reliable machine, a deterrent to theft, a statement against consumerism, and my ticket to adventure. I liked the look.

Stuffed into the bags that were clipped, tied, and fastened to my bike was a collection of gear, old and new, that I needed to make the trip. Over my rear wheel, a rack held two cat litter containers I had turned into homemade bike panniers. Those buckets contained a fleece jacket, raingear, a pack towel, shower supplies, tools for minor repairs, a water-color set, two cooking pots, one homemade stove, one day's worth of food, a bike lock, and a large water bottle. On top of the buckets were my tent, a folding chair, and a tripod, all held in place by bungee cords and a sign announcing my route and website. One side of the sign was in English, the other in Spanish.

A rack over the front wheel held two store-bought red panniers. One contained my sleeping bag, journal, book, and headlamp; the other, my rolled-up air mattress, laptop computer, and charging devices. On my han-dlebars was a small bag, stuffed with my camera, phone, wallet, passport, maps, sunscreen, toothbrush, spoon, and pocketknife. It all added up to something around seventy pounds. In contrast, each monarch weighed half a gram. It takes about four monarchs to equal the weight of a dime. Though people gasped when I told them what I was doing, it seemed to me that the monarchs, with their unburdened wings, deserved the accolades. They were much better-equipped adventurers than me.

Bags packed, I waved goodbye to all the friends I had made while wait-ing in Mexico and pushed off. The first breath I took filled more than my lungs. After months of planning and anticipation, I was finally on my way. There was nothing left to do but do.

Gravity carried me downward. I went fast, weaving around potholes, finding gaps in the speedbumps, and feeling the excitement of the start. Suddenly, with no warning, I went from enjoying the views to colliding at full velocity with a classic, hard-to-spot, Michoacán speed bump—better described as a moderately high curb stretching across the road. In what seemed like slow motion, I catapulted into the air, bringing my bike along thanks to a death grip on my handlebars. The ground rushed forward and I strategically braced for the whiplash of a jarring landing.

THUMP!

My wheel hit the ground, I hit my brakes, and one of my panniers was sent flying with the jolt. It hit the ground without grace and unpacked itself in the middle of the road. I laughed sheepishly, looking to see if anyone had witnessed my blunder. "Off to a good start," I said to the empty road. At least I was on my feet. I repacked and resumed.

Reaching the bottom of the hill, I turned and started trudging right back up the same mountain on a different, more deteriorated road. I stumbled upward as the road became a mess of rock and red dust. At the sections with the loosest rubble, I was forced to jump off my bike and push. As I walked, I glanced upward. Streaks of monarchs hung above the road. Like a sign, they guided me; I became a witness to a pilgrimage of wings.

The crowd of monarchs grew from a trickle to a river. When the road turned, the current of orange gathered me in its arms. I matched the speed of the pumping and gliding wings. Neither hurried nor lazy, we could travel as one. Against a backdrop of dirt road, monarchs floated below me. Against a canvas of blue sky, monarchs soared above.

The butterflies flew, and since I couldn't fly, I biked. After so much planning, so much dreaming, so much work, I was officially butterbiking with the butterflies. The name of my project, Butterbike, finally made sense.

Then the butterflies ditched me.

"Monarchs!" I hollered, knowing exactly how absurd I sounded. "Come back!"

Since they didn't have to follow roads, they confidently cut through the forest on a direct route north. Following them was impossible on my bike, so I knew I would have to traverse back and forth across their path on whatever roads I could find. I wished the migrants good luck as I turned away with the bend of the road, and we headed into our divergent unknowns.

I was less capable than the monarchs of navigating the unfamiliar, and I proved it at the trip's first intersection. When the road forked, I was presented with two choices: follow the curve of the road to the right, or turn left. Both dipped downward and out of sight.

Without a smartphone, my only option was to rely on the clues of the road, my horrible sense of direction, and my paper map. This map had been only semi-reliable on my first bicycle trip through Mexico. The creases were worn and I'd covered most of it in clear tape to protect it. What I was protecting, however, was a colorful piece of paper that was accurate about 70 percent of the time. Even if the cities were misnamed and some of the roads didn't actually exist, I used it because it was better than nothing.

According to said map, I had to leave the sanctuaries, go east toward a town with a name I couldn't pronounce, and take the first left after crossing from the state of Michoacán to the state of Mexico. Easy—right?

Standing at the first junction after crossing the state line, I double checked my map and chose the left-turn option. Unsure of my decision, I rode hesitantly, gripping my brakes as my bike tiptoed downhill. The slow passing of the trees lining the road reminded me I was missing an opportunity. Braking going downhill is like watching a bowl of chocolate ice cream melt rather than eating it. Basically, unforgivable.

I let go of the brakes, committing to the consequences of my navigational guess, and tucked my body into itself. Gaining speed, the trees blurred through my watery eyes, and the wind drowned out my worries. At forty miles per hour, I thought about nothing but the road ahead, analyzing each shadow for hidden potholes, speedbumps, and bowls of chocolate ice cream (you never know, right?).

Going downhill fast is my version of flying.

At the bottom of the hill, the thrill of flying was replaced by the realization that at the first intersection of the trip—literally, my first opportunity to go the wrong way—I had gone the wrong way. Confirmation of my error came with the passing of several miles, and the absence of a junction to mark the next turn. I knew where I wasn't. I didn't know where I was. I was officially lost.

When lost, one has two options: turn around or keep going. You can't stand still and do nothing. Backtracking sounded torturous, so I chose to keep going. I would carry on, find a sign or a human that could tell me where I was, then revise my plan from there. Decision made, I set off, ready to accept whatever I found. Second guessing can become toxic.

Fortified by having made my first blunder, and on my way to solving it, I felt another rush of freedom. Mistakes are less scary once you have made a few. It was clear, however, as the clouds blushed with the first sign of setting sun, that I was not going to completely solve my wrong turn that day. I would need to camp, and that was fine by me. Nothing solves problems like escaping into a tent. In the morning, fresh from sleep, I could trace a new route north.

My camping options were not obvious. There were open fields of young corn, rows of spiderlike agave plants, clusters of colorful cement houses, and the occasional grove of spared pine trees. Even though I had biked thousands of miles and deliberated over camping spots hundreds of times, each night was its own puzzle.

I slowed my bike as I passed one of the mentioned options: a cluster of trees just off the road. It wasn't perfect, but the sun had flipped the switch and the countdown was on; unless I wanted to pick a spot in the soon-to-be freezing darkness, I had to settle.

A small path just off the road took me to a flattish spot under a tree. *Home sweet home*, I thought as I let the weight of my bike be absorbed by the ground. Voices, mumbles of Spanish from the people walking down the road, filtered through the trees. Since I couldn't see them, I assumed they couldn't see me. I wasn't scared of them, but I felt more comfortable knowing I was well hidden. My tent fit perfectly on a carpet of pine needles, and even with all my sleeping gear thrown inside, there was plenty of room to spare. It was my first long solo trip, and if nothing else, it would be nice to have a roomy tent, practically a mansion.

Two bites before finishing my sandwich, a version of dinner I had elected because cooking required more energy than I could muster, a man's whistle and the trotting footsteps of a horse broke the silence. I could neither hide my tent, nor try to hide in my tent, so I paused to listen in the near dark.

The horse and man walked ten feet from me, on a faint trail I hadn't noticed before. He was likely commuting between his fields and his house and would never have guessed that a woman from another country had stopped to sleep in his neighborhood. I wasn't sure if he had spotted me, but I was sure I didn't want to startle him.

My best Spanish greeting broke the silence. "Buenas noches."

I took his silence for a question and followed up with a choppy explanation of what I was doing. He looked at me, through the dark, but didn't stop. Perhaps he understood me, perhaps he didn't. Either way, he didn't smile or talk. He barely reacted. The only indication that I was real came from his eyes, which followed me until a low branch forced him to duck as he slipped into the night. Discovered, I had two options, pack up and find a better spot, or go to bed. I went to bed.

When the new day's sun slid across the forest floor, any worries I'd had about my camp spot were extinguished. I welcomed the second day of my trip. *Happy Birthday*, I told myself. Perhaps if I had known I'd be celebrating the big three-two by biking down a terrifyingly narrow highway full of terrifyingly fast cars, I would have turned around, found a cake shop, and called it a day. Instead, I moved forward. Like a video in rewind, I repacked all my things and reloaded them onto my bike. The spot I had turned into a home for the night lay behind me, just as I had found it. Except for some flattened pine needles, no one would ever know I had been there. It was satisfying to borrow a random place, make it my home, take care of it, then return it as close to untouched as possible.

Reversing course, I headed back through the trees and picked up where I had left off. A few miles later, I reached my first junction of the day and turned right. It took only moments for the first car to pass me. I cursed as its wake sucked me and my bike toward the middle of the road. I braced, cursed some more, and kept going. Car after car, my stress grew. Mile after mile, my doubt grew. "This is not fun," I told the thousands of miles of road that extended like a paved plague before me. It was only Day Two. *What had I committed myself to?*

Doubt is as much of an adversary on a long trip as tired muscles are. However, just as legs can be conditioned to carry one farther, a mind can be conditioned, too. The key, at least for me, was to ignore the big picture. Never project thousands of miles into the future. Instead, think about the next mile, the next town, or (best of all) the next meal. In this way, I could confront small distances, and celebrate strings of tiny victories that would soon add up. I knew this strategy because I was not on my first long trip. I had already pedaled thousands of miles, including a twelve-country bicycle trip from Bolivia to Texas and a forty-nine-state tour around the United States. What these trips had in common was the sense of impossibility that lingered at the start. Before each trip, people told me my dream was not attainable, that I would probably die. Before each trip, I worried that I would fail. But by continuing, I had proved each time that a mile is a mile, regardless of how many are strung together.

I celebrated my birthday and my survival by ending my day long before the sun did. After sixty-five car-crazed miles, my mind, legs, and butt were all pleading, *STOP!*

An abandoned road ran like a shadow alongside the newer highway. Old roads can make great campsites, and this one was no exception. Blocked to cars by a few boulders, the road extended out of sight. It had a gradual slope, but I knew that I could stuff my rain jacket under my air mattress, like a retaining wall, and turn the hill into something resembling flat. With some loose rocks, I staked my tent on the pavement where cars had once sped. Sitting on the centerline, I ate another sandwich. Dessert was a birthday chocolate, a gift from Brianda.

Following Mountains

DAYS 3-5 / MARCH 14-16

MILES 118-310

Although the exact route of the monarchs from their overwintering sanctuaries in central Mexico to the Texas border is not fully understood, scientists hypothesize that the butterflies use Mexico's mountain ranges as their guides. The Transverse Neovolcanic Belt of mountains, cradling the oyamel firs and their winter tenants, is the first shepherd of the journey. Its long arm runs east–west like a belt cinching the narrow waste of Mexico. The monarchs trace its volcanic memory east to a junction with the mountains of the Sierra Madre Oriental, which run north–south and divide Mexico's dry center from its green eastern edge. The veering monarchs use these mountainous cues, like a sight line, to follow spring north.

To match the monarchs I consulted my maps, but found few roads that followed the mountains. On the east side, where water invited development, the domination of varied topography pulled the main highways far from the ridge and nearly to sea level. To the west, the desert loomed like a guard dog, intimidating towns, roads, and me. On both sides, any road was at the mercy of the mountains, which wove through valleys, climbed up and over passes, retreated from obstruction, and were superfluous in their mileages. Unable to skim the spines of the ranges and flutter from point to point like the butterflies, I was relegated to the roads.

My immediate route, neither graceful nor direct, became a zigzagging course about which I was often unsure. Some byways rambled toward peaks where I imagined skies of monarchs lifted out of reach. Some roads headed toward Texas, where milkweed seeds were awakening to the gentle nudge of a young spring. These pathways, scribbled on my map, at first were only lines. Two-dimensional marks on a two-dimensional map, trying to explain a three-dimensional world. Only the map's rainbow background hinted at just how flat the world isn't. Every thousand feet of elevation had its own color, from the green of sea level to the brown of a mountain vantage. The colors through which my route traversed warned me of an impending climb.

Noting the climb and the menacing heat where it began, I camped in the dry belly of a lonely valley and set my alarm. In the cool of the rising sun, I would climb toward the sky.

When my alarm rang out, I forced myself to begin the day. Arrogant birds sung tributes to the fresh day, strange plants waved spines in fireworks-like explosions, and red flowers hung from gnarly limbs like sparks of confetti. My belongings rolled, folded, stuffed, and loaded, I aimed my bicycle ceremoniously at the mountainous wall I planned to scale. It was still sleeping, tucked into a bed of clouds.

I climbed uphill for three hours, on a steady-but-not-steep road. The layers of mountains unfolded gradually. The views distracted me, as did the roadside vegetation, which seemed to transform with each hairpin turn. Plants adapted to higher elevations replaced the lower elevation plant species, as the traffic dwindled into a peaceful quiet. Aside from the occasional car, I was left to ride unencumbered by worry or belching smoke.

At the top, I celebrated with a mango snack and silly photo shoot. Then, beginning the downhill payoff, I coasted around a corner and realized that I had been celebrating ignorance. Before me lay not the glorious downhill fun I had predicted. My future twisted upward and disappeared around the mountain's high shoulder.

Three more hours of climbing drained my water bottles, the last of my food, and the power from my legs. The road became a portal, from relentless heat to cloud-soaked forest. The mountainside grew greener, the red embers and yellow arcs of roadside flowers grew thicker, and tilled lands grew more common. The dry desert was left behind. All the while, my back flickered with pain, my legs burned, and my eyes traced the road ahead, desperate to see a high point that would mean a subsequent descent.

Seven hours into the near-constant climb, a truck pulled off the road just in front of me, and a smiling man leaned out the window. I watched, ready for anything. The man stepped out. In his hand was a heavy bag of peanuts, which he held out for me. As luck would have it, Ramiro was a cyclist, with high-calorie snacks and intel on the road. "Thirty minutes to the top," he told me, as I munched on peanuts. I was certain that he was overestimating my speed. Most people couldn't imagine my snail's pace on an all-day climb with a fully loaded bike. Thirty minutes to him was likely an hour more for me.

Ramiro would soon learn exactly how slow I was, because as I ate peanuts he proposed meeting up and biking for a few days together. Happy to have company, I said yes before turning back to the hill at hand.

An hour later, the steep climbing relented and gave way to a gentler grade. The road eased its way through a thick frame of silvery pines, heavy with drooping needles. Beyond the roadside trees, a rainbow of greens sprang from the unseen mountain floor, covering all but the white rock scars of the twisted mountains. A thin veil of low-lying clouds spread like strokes of an eraser.

I was 6200 feet higher than my previous night's camp spot when I finally arrived at the crest of the road, where views of both sides of the mountain met. As a reward, I propped my bike against a retaining wall, and stretched my spent legs with an extra-slow walk. Wisps of clouds carried shivers through my sweat-soaked clothes. I donned my fleece and rain jacket, got on my bike, turned my back to the low-hanging sun, and headed down the other side of the pass.

The descent, as with so many, started like a sigh. My muscles rested while gravity reigned. The break was a relief, like taking off a heavy backpack or entering a warm building after a long, cold stint outside. As the wind began to scream, I tucked in like a racing pro, and let the sound of my tires, linking road to sky with a thin trail of rubber, hum beneath me.

I hit my fastest speed on a straightaway as pieced-together houses, fences of feral plants, and generations of people living on the mountain's slopes blurred by out of the corners of my eyes. I didn't try to catalog what I saw and focused on the thrill of flying. To feel the speed was to feel alive, feel the air above and below, feel the distance come and go, feel the risk of the smallest mistake, feel the stillness in a spinning world. As my elevation dropped and I traded hours of climbing for minutes of flight, I felt the temperature climb. It was not the dry heat I had left behind, but a rich, moist jungle heat. I leaned into each turn as the road and I were swallowed by a forest of clouds, green, warmth, sound, and a monarch.

A monarch!

I stopped as fast as one can at forty-five miles per hour, just in time to watch the monarch circle overhead. From its black veins carved in orange to its black-and-white-speckled edging, there was no doubting my identification. I abandoned my bike on the ground and started running after the butterfly. It was the first confirmed monarch I had seen since leaving their overwintering grounds. As it flapped goodbye, I turned from chasing to celebrating. I unleashed my happiness with a frantic flailing I would later describe as my "monarch happy dance."

Cars swerved to distance themselves from me as I twirled like an out-of-control puppet. Had they stopped to ask what I was doing, I would have told them that I was commemorating the vanquishing of a fear. Since announcing my trip, the slight pestering of doubt that I would not see a monarch on my route had harassed my mind.

This one's presence confirmed I was on the route of a great migration.

I met back up with Ramiro, the cyclist who had offered me peanuts the previous day, in the square of a tucked-away town. Together we spent the warm, sunny afternoon tackling dirt road climbs on an adventurous off-highway route. His quick pace down the hills motivated me to be more daring, and I put my trust in my bike as I skidded around corners and braced for the bumps of each dusty descent. On the ascents, I was slower, my bike weighed a lot more, but Ramiro didn't mind waiting. By the end of the first day, we were becoming fast friends, kindred spirits. In the cement yard of a school, we set up camp, bought a home-cooked meal from a family living nearby, and talked to each other and the stars until we couldn't keep our eyes open.

The next day unfolded much the same. On a long uphill climb, I soon lost sight of Ramiro. I stopped to photograph the growing mountain horizon and its intriguing canyons. I traced its peaks with my eyes and imagined diving into the strange pockmark of a cave on the mountain's side. When I reached the main highway, I found Ramiro drinking his second cold soda. All that was left was to go for a swim in the emerald-green river that cut the valley in half. That and to fill my water bottles.

In Mexico, the water from residential faucets is typically not safe to drink. Instead, people buy *garrafones* of water: five-gallon, blue plastic containers that can be refilled with filtered water. Unwilling to buy disposable plastic bottles, I would find a store, restaurant, or house with a garrafón and ask to fill up. I always offered a few pesos in exchange; people were happy to help (and rarely accepted my money).

I gathered my three metal bottles and told Ramiro my plan, to which he responded, "No." He said I should drink water from the river when we got there.

What? I was well versed in drinking water straight from the earth. I took great pleasure in filling my bottle from a mountain stream or sipping at the surface of an alpine lake as I swam. At El Rosario, hoses laced the mountains and brought fresh, clean water to people's houses. When thirsty, I only needed to find a hose. Each gulp of wild water was a gift, a resource more valuable than gold. And yet, I would never drink straight from a large river fed from the unknown. Humans long ago squandered the purity of most water. I countered his no with my own.

I realized he had suddenly become very serious. He didn't want me drinking water from a garrafón. It was river water or no water at all. I was shocked that he was so upset.

"You are not listening," he snapped. I ignored him, asking a store owner if he had a garrafón. "You never listen!" he screamed. I can't remember if he spoke in English or Spanish, as we had developed our own brand of Spanglish, but his loud "never," referencing the one day we had known each other, had my attention. I was aware of the escalating tension, but I was too indignant to ask for an explanation of his anger or to back down.

My stubbornness comes in great for adventures. When others might give up, I dig in and, regardless of the discomfort, I am there for the long run. I thrive on it. This trait doesn't translate as well with people. With people, there needs to be a level of give that I rarely can summon. I stand ground that isn't mine. I don't let things go. As he got angrier and angrier about my refusal to drink from a river I had never seen, my stubbornness took charge.

English and Spanish intensified and our argument boiled into a fight. He stepped forward, narrowing the gap between our angry faces. I imagined him punching me, and at that moment I didn't care. I wasn't going to let him intimidate me. Instead of stepping back and de-escalating, I stepped forward.

It could have been the cleanest river in the world, but I wasn't going to risk it. With our faces inches apart, as if in a cartoon yelling match, the air became clogged. Our words gained volume as they lost substance. I wasn't going to change his mind, so I fell into silence. I had no idea why he had turned so quickly into a raving tyrant, so all I could do was leave on my bike without another word. Half scared, half angry, I pounded out the miles between us. I felt as if I had actually been punched, and to keep from crying I put all my energy into moving forward. Without water, I was thirsty. Without a swim, I was unrefreshed. Only distance mattered.

It took twenty thirsty miles to reach the next small town, and I marked my arrival by drinking two glass bottles of icy soda in the shade of the store. The physical effort of the ride had drained my anger, and I was left with the confused aftertaste of confrontation. I dissected the moment scene by scene, trying to write a version where I had found the strength to remain calm. All I found was the truth, which I wanted to let go. I filled up my bottles, drank them, and refilled them from the store's garrafón. The store owner's kindness and refusal to let me pay for the water helped soften the edges of my bad memory. I ate a few mangoes, a few sandwiches, and drank one more deliciously cold soda (too much soda, I know).

Many travelers carry burdens with them, burdens they hope to escape by constantly moving. Many travelers know that while this doesn't usually work, sometimes it does. I left on my bike feeling strong.

Deserted Miles and Trials

MILES 310–422

Knuckles white and teeth rattling, I trained my eyes on the dirt directly in front of my wheel. My progress was tedious, the rocky substrate unforgiving. Only when coasting was smooth could I steal a glance at what lay beyond. What I saw worried me. I took a deep breath, riding toward the belly of the beast and bracing for some frustrating news. *I'll laugh about this someday*, I desperately reminded myself as I came to a stop at the end of a dead-end road. A gang of grazing goats and cows loitered.

The wrong turn had come after a stint of perfect biking. I had left Ramiro and his aggression behind, and was celebrating my route choice. Mild hills—troughs of semiarid scrubland and crests of oak woodland—crafted leisurely days and acceptable mile totals. Crowds of cacti with craning necks like telephone poles, and tunnels of oaks with mossy beards and arthritic limbs cradled narrow roads. The miles were unencumbered by traffic. Each day was a surprise, revealed by the spinning of bicycle wheels. Each night was a surprise, revealed by an invitation from providence. Where I camped was a matter of where I arrived.

One night, I camped at the far end of a soccer field, where kids stopped their game to circle my tent and watch me cook. I shared my frog-shaped chocolates with them as my pasta boiled, and I wondered if their families would believe the story they would later recount. The next night I set up my tent in a courtyard guarded by the castle-like walls of the Protección Civil, a nationwide administration that provides disaster assistance. While armed guards smoked cigarettes, I fished water from a well and washed my underwear in the only vessel I could find: my cooking pot (obviously disgusting, but so it goes). The next night, I pushed off the road, and camped in the company of the wild.

Bike, camp, repeat. That was the pattern of my first days and weeks, made from connecting unpredictable and often unimaginable dots. Tough days and random nights quilted a fabric of adventure.

My dead-end blunder was one such adventure, though it was not entirely unpredictable. The officials at the Protección Civil had warned me against my route. They had insisted I travel the main highway, a boring runway of pavement clogged by big semis. Their hospitality obligated me to listen to their advice, though that didn't mean I had to follow it. As they told me where to go I nodded vehemently, pretended to be in agreement, waved goodbye, then headed away in the opposite direction. Now that I know, I would still follow my own lead.

For about twenty-five miles I couldn't believe how ridiculous their force-fed counsel had been. Unlike the highway they recommended, the paved road I'd taken had only a trickle of traffic and was flanked by a string of quaint towns. I was surrounded by a swarm of colorful butterflies and the flowers they courted. When the road turned to dirt, I was happier than ever that I had ignored their warnings.

The dirt road kept my mind from wandering, by forcing my attention on picking a line and navigating rough patches. There was a thrill in dodging landmines of jarring rock, finding the smooth runs fringing the road, and picking up speed as gravel blurred and concentration focused. Adventure and purpose sprang from difficulty.

As the road climbed steadily it also deteriorated, finally shooting up like a decaying geologic wall. The jumbled chaos of rock, generously called a road, forced me to dismount and walk. Still, I didn't regret my route choice. Ribbons, like prayer flags, flapped in the wind—a multicolored spider web radiating from a small shrine for the Virgin Mary. I was on a pilgrim's path, a middle-of-nowhere journey. I was on an *adventure*. At the top, an immediate descent, nearly as steep as what I had just climbed, greeted me. I whipped down, jolting and jerking. As the valley below revealed itself, so did the truth. The road faded into a flat expanse of grass encircling a nearly evaporated lake and fenced by a dotted line of houses. I scanned the encompassing hills with growing dread, desperate for a continuation of the road. "Hummm," I declared, sarcastic and suspicious, "this doesn't seem right." To confirm my hunch, I continued down what was left of the road.

A goat raised its head, mirroring my confused look. There was no denying it, I was at the "end" part of a true dead-end. The road had completely dissolved into grass, so I turned, passing the unconcerned grazers, and headed to the closest home. I stopped at a fence and waved at a woman on the other side. She did a very good job of masking her surprise, which must have been considerable at encountering a white lady on a bike, blabbering in strange Spanish, about a road that was obviously not a road. She confirmed the obvious: the road on which I had arrived was the only road on which I could leave. Pointing, she patiently gave me directions: go back up the hill, back down the other side, and take the first right.

With resignation, I climbed back up what I had just come down, the questioning stares of the entire village burning through my back. The disappointment of having to backtrack festered as I completed the tedious U-turn. Sullen, I calculated the hours lost, muscles taxed, calories burned, and water gulped.

The stares continued as I descended the same hill I had trudged up hours earlier. Swallowing my pride, I asked curious onlookers if I was going the right way. The lady outside her house, two kids on bikes, a couple on a motorcycle, and a guy loitering with a donkey all confirmed my revised yet familiar route.

At the first right-hand turn, it became very obvious why I had missed it on my first pass and why the guys back at the Protección Civil had been so committed to their warning: the road was not really a road at all. "You've got to be kidding me," I said to the bleak hills toward which the faint track led. It was not a road. It was more of a cow trail. This was the route I had insisted on, though—the route that the officers had tried to dissuade me from, the route I had missed, and the route on which I would now continue.

With a big sigh, a shrug, and a *Here goes nothing*, I committed to the track, heading into the unknown. At least I was no longer backtracking.

The trail was twisted by hills that at first were stubbled with scratchy creosote and later bearded with a spiky mat of yuccas. I followed it downward at a meager pace. Gravity wanted me to go faster, but the path, a ribbon of talus, forced a battered crawl.

That crude road was the only sign that humans had ever passed through such desolation. It gave company to the yuccas that guarded the vastness with swordlike leaves erupting haphazardly from their scaled trunks. From road to horizon they twisted, less like plants and more like spiky haired dancers frozen mid-move in a tango with the heat, the wind, and the occasional butterfly. Curious and ready for a break, I abandoned my bike and the road to walk among them. I stretched my arms and extended my fingers to mirror their convoluted poses.

Back on my bike, the road was nearly mine alone. The only other people I saw, two guys driving up the road, were so surprised to find someone else that they stopped their car and asked to take photos. Later I would laugh at their timing. They thought *biking* the road was unbelievable. Had they driven by ten minutes later, they would have found me on the side of the road, howling at the yuccas, doubled over in pain; the unfortunate result of pausing to inspect a beetle printing whimsical tracks in the sand.

I had seen the beetle from my bike and stopped to follow it. Absorbed in the details of the desert floor, I leaned in. Like a needle, the dagger-like tentacle of a nearby yucca penetrated my bicep. I roared in painful surprise. No one saw me as I flailed about, a tortured, bizarre tornado blocking the road. Had those men driven by then, instead of taking photos, I'm sure they would have slammed on the gas and fled. I smiled at the thought, once the shooting pain had faded into a distant pulsing echo (my arm

50

would still hurt, three weeks later). To travel is to live alongside uncontrollable timing.

As if the world knew that I had been having a tough go of it—that I had been cornered by a dead-end, slowed by bumpy roads, attacked by yuccas, and was currently battling a growing headwind—it sent a monarch to my rescue. The butterfly charged through the headwind, gripped the feathery leaves of a roadside acacia, and said with its presence, "Follow me." Beyond the wind-prodded plants lay a seemingly barren wasteland of worn-out, white land. I had been expecting to see a monarch about as much as I had expected to find a bowl of still-frozen ice cream. The monarch became an arrow pointing me forward. I may have been struggling, but at least I was struggling on the route of a monarch—though it seemed as out of place as I was.

That motivational monarch was only the third I had seen in eight days of riding, and while my expectations were low, the fear of missing them was real. Imaginary roads full of monarchs spun in my head. I told myself to be patient and look at the real goal: to connect with the people who could save them.

My travels demonstrated my passion. I had committed 10,000 miles' worth of devotion to the monarchs. I hoped that my wonderment would catch on, that pride and a sense of responsibility would follow. My ride was to be a conversation starter, an invitation into the monarchs' world. I strove to be a wingman, helping people fall in love with the monarchs the way I had.

In the United States and Canada, I was organizing more formal presentations at schools and nature centers, but in Mexico, with only a few weeks of stories and the vocabulary of a three-year-old, I was satisfied talking to small groups. When I bought a Corona beer (it was the only thing cold enough to be refreshing), I sat in the shade of the store's porch with two wrinkled cowboys and talked about monarchs. When kids ran alongside my bike, I told them about monarchs. When I sat eating *gorditas* at a town square, I told the women cooking about the monarchs.

Almost everyone I talked to in Mexico said they knew about monarch butterflies, and with an unambiguous consistency they told me that the butterflies only passed through in the fall. Since monarchs fly north in the spring and south in the fall, this observation needed explanation.

My first thought was that there are fewer monarchs to notice in the spring than in the fall, because the spring population is decreased by winter's many hazards. Three such overwintering hazards include black-backed orioles, black-headed grosbeaks, and black-eared mice. Unlike most vertebrates—which cannot eat monarchs because the butter-flies have a heart-stopping cardenolide toxin that induces vomiting—these three species have evolved strategies to deal with the toxin and take advantage of the abundant fat reserves the monarchs embody as they hang like grapes in a vineyard.

In the mornings and evenings, when monarchs are too cold to escape, hungry black-backed orioles (*Icterus abeillei*) feed. With orange breasts, black backs, and white highlights, their feathered cloaks blend perfectly with their scaled prey. Sharp beaks split open monarch abdomens and eat the soft, fatty insides. This strategy avoids the monarch's cuticle skin, where cardenolide toxins are typically concentrated. It's a method similar to that of a kid avoiding the crust and eating only the soft parts of a PB&J sandwich.

The black-headed grosbeaks (*Pheucticus melanocephalus*)—painted similarly to black-backed orioles, but with a stout bill—instead eat the monarch's entire abdomen. Such victims are easy to identify as they float down from the canopy. Abdomenless, their wings still beat in protest. Perhaps because grosbeaks eat the toxic cuticle, they tend to prefer male monarchs, which can be up to 30 percent less toxic than females.

Of the thirty-seven bird species in the overwintering area that eat insects, only the black-backed orioles and black-headed grosbeaks consistently eat monarchs, likely around 15 percent of the population each winter. Studies at the sanctuaries have found a high degree of variation between mortality rates at different colonies, a likely result of canopy intactness and colony size. When monarchs cluster in areas of the forest where the canopy is less intact because of logging, disease, wind damage, or fire, birds can more easily access and consume high numbers of monarchs. Furthermore, clusters tend to be smaller at less-populated colonies, which means that a higher ratio of monarchs rest on the exposed outer edge than in the protected, hidden center of each mass. It seems that smaller colonies, in disturbed forests, are the most vulnerable to bird predation, where flocks of five to sixty can pick them off one by one.

At night, monarchs face a different predator: the black-eared mouse (*Peromyscus melanotis*). Of the four species of mice in the oyamel forest, only the black-eared mouse has adapted to eat the toxic monarchs. The doe-eyed, clown-eared mice target butterflies on the ground (another reason besides temperature for monarchs to crawl onto vegetation even slightly above the ground), and researcher Karen Oberhauser has observed them eating as many as thirty-seven monarchs a night. Piled monarchs, with only their wings remaining, are the calling cards of these mice. The wings, with a high concentration of toxin but no fatty reward, are not worth the effort.

Predation, along with winter's other burdens, results in decreasing populations without any breeding to replace the losses. Come spring, though, the winter survivors fly north to Texas, where they encounter milkweed and lay the next generation. Each female typically lays between 300 and 500 eggs (one captive female was recorded to have laid 1179 eggs), and if only 1 percent survive, that will still double or triple the population arriving in Mexico. All summer, as one generation after another lays millions of eggs, the population rebounds.

Another explanation for the butterflies being less noticed in the spring might be seasonal behaviors. In the spring, the monarchs prioritize travel over feeding. Their goal is to arrive in Texas and lay eggs, and to do this they fly high and search for wind currents. When wind direction is unfavorable, monarchs will fly low to the ground, but otherwise they stay high. Reportedly, the highest recorded monarch was seen at 11,000 feet by a glider pilot. Any monarchs flying higher than 300 feet off the ground are invisible to the naked eye, so high-altitude spring migrants would definitely be less noticeable. In the fall, however, the monarchs' goal is to sip as much nectar as possible and gain fat reserves for the winter. To do this, they spend much of their time gorging on flowers at human level—and being noticed.

Though I favored my first two theories, I added other possibilities, including the hypothoses that the monarchs have different northbound and southbound routes; that fall winds drive more stragglers to the dry western slopes; that people are outside more in the fall than in the spring; and simply that it was a matter of misidentification. Maybe it is a combination, or something entirely different.

While the monarch is one of the most studied insects in the world, it still whispers in a secret language. Though I knew I could never understand it all, I hoped that by continuing on, some answers would reveal themselves. I also hoped no more yuccas would stab me.

Ice Cream and Tacos

For me, the best bike tours alternate between dirt and paved roads. Dirt roads are slow, demand concentration, cut through the less known, and invite a difficulty that can turn a cyclist into a pioneer. Paved roads are efficient, permit the mind to wander, connect to more services, and make a cyclist feel like they are getting somewhere. After a few days on one, I crave the other.

By the time the dirt road hit pavement, I was barely wobbling north—an exhausted, yucca-stabbed survivor. I celebrated the switch by coasting comfortably on the buttery smooth pavement. For the first time in days I felt luxuriously fast, and this relief carried me to Tula, Tamaulipas. There I paused for a night with a wonderful family I met on the road. Eating bean soup from a Styrofoam cup, I recounted some of my best stories from the road. What amazed them most, though, was how heavy my bike was. We giggled when none of them could lift it off the ground. The next day I carried on to the main highway and its flurry of traffic. In the ribbon of pavement shouldering the road, I concentrated on pounding out miles and lulled myself into a trance.

I didn't notice the speck of a motorcycle in the distance, hidden among the humongous trucks and laboring cars. I didn't notice that as it approached, it slowed. I only looked up when it pulled up directly beside me. Suspicious, I took stock of my situation. *What does he want?* I wondered as I stopped my bike, aware how far from anywhere I was, and surveyed the man. He wore a floppy, blue fisherman's hat, yellow safety glasses, and a windbreaker. Tied to his handlebars were plastic bags, and tied to the back of his motorcycle was a red plastic cooler.

Soon it became obvious what he wanted.

Later, when telling this story to groups, I would always pause here to drag out the audience's worst fears—and maybe shine a light on our biases.

With a wave of his hand, he directed me to the cooler, and offered me an ice cream cone.

It was the perfect treat for the hot afternoon. My bike computer declared the temperature at nearly 100 degrees F. The strawberry drips running down the pink cone confirmed the thermometer's accuracy.

Every day I met people interested in my trip, and every day those people became part of it. Their kind words, refreshing snacks, and inspiring hospitality were a tailwind, helping me forward. More than calories, that ice cream was a gift of support, a reminder that I was part of a team. I told the man about the monarchs, and he told me that we can trust the good in the world.

Then the man asked for my hand in marriage. "No," I told him, "but thanks for the ice cream."

I flipped between patience and frustration, the latter a feeling that was growing familiar. Getting stopped to answer predictable questions, often ten or twenty times a day, drained me. The interruptions killed my momentum, the entitled curiosity got under my skin. It was hot, I was tired, and sometimes I just wanted to get where I was going. I was disappointed by my petty irritations.

Soon after the ice cream stop, my frustration came to a head and I was forced to examine my irritability. On a cobblestone climb, I encountered an assembly of students in matching school uniforms. All of us were headed uphill, and the steepness and difficult road surface meant I was moving only slightly faster than they were as they walked toward their school. One after another, observant students would realize I was nearby, then abruptly avoid my smile by turning away with a jolt. In a huddle, they would whisper, then turn to sneak peeks at me. I'd say, "Buenos dias!" to break the ice, but they would just look at me as if I had said, "I am a cat."

It felt degrading and annoying to be simultaneously ogled and ignored. I am not proud of it, but their game became mine. On my slow, upward crawl, I watched a gaggle of girls whisper, and just when one turned to get a conspicuously inconspicuous look, I caught her eyes, smiled, waved, and answered her silence with a "BOO!" She turned back so fast, it was like I had burned her. My shame at my behavior made the climb steeper.

My reaction scared me. The point of my trip was to bridge people and monarchs. Staring was a start, and if someone wanted to stop me with questions, wasn't that the point? I wrote, "It is easier to be nice" on a piece of paper that I tucked in my handlebar bag.

It was not long before I was stopped and, like clockwork, a man in a car asked me the common questions of where I was going and where I was from. Then he asked me the most common of the common questions.

"Estas sola?"

Are you alone? It was never threatening, more disbelief. For some reason, the question always grated on my nerves. It was the assumption that a woman couldn't handle doing anything on her own. Instead of being curt, I decided to practice patience.

"Si," I said with a smile.

"It is just you?" he asked again.

"Yes," I repeated. This question was always followed up by its many variations, like a coping mechanism for disbelief.

Solo?

By yourself?

No one else is with you?

Yes, yes, yes. I answered each iteration, trying my best to remain pleasant while squeezing in bits of information about my mission.

Finally satisfied, the man drove off, knowing more about monarchs and biking than he likely ever would have had I not stopped.

I took out the slip of paper scribbled with my advice and added, "Do it for the monarchs." The questions and my answers could connect people to butterflies. It was the reminder I needed, even if it didn't always work.

I knew, even then, that my exhaustion was a major catalyst for my unchecked attitude. The long days and unrelentingly tough cycling were zapping my strength. I was giving it everything I had, and I could barely keep my average speed above ten kilometers per hour. Not ten *miles* per hour like my normal touring speed, but ten kilometers per hour, which was six miles per hour. Such slow speed meant I had to bike all day if I wanted to reach my sixty-mile-per-day goal and bank miles for a future rest day. Canada seemed impossibly far, doubt seemed extra persistent, and perhaps I was taking my frustration out on innocent people.

Following my own advice paid off a day later, when I stopped at a small store in an equally small town after a grueling battle with wind and gravity. The store was my escape. I ducked out of the wind and into the stillness of shelves stocked with supplies. The first thing I noticed was the store's above-average cleanliness and meticulous organization.

As previously mentioned, nearly every house in the countryside of Mexico includes a store, which is usually attached to the house. There is usually a corner with fresh tomatoes and potatoes mingling with their overripe or wrinkled counterparts. Stacks of crated eggs, noisily humming refrigerators with cold drinks, bags of beans, and cans of chilies were all dependable finds, I had discovered. I spent a lot of time in these stores escaping the heat with ice-cold sodas sold in returnable glass bottles. I also stocked up on snacks, and if I was lucky to find them, I treated myself to mangoes and my favorite mint chocolate-ball candies.

Vicky, the woman behind the counter, smiled as I remarked on the quality of her store. After buying a soda, I sat on a stool and answered her questions. Remembering my goal, I did my best to include a smile, patience, and a peppering of monarch facts as I explained what I was doing.

"Quieres tacos?" Her next question, inviting me for tacos, had only one answer.

"Si!" I replied enthusiastically.

Saying yes is as important on a bike tour as drinking water or packing a toothbrush. It's part trusting strangers, part trusting yourself, and part taking things as they come. Saying yes was a skill I had been honing since my first long bike trip.

I had been surprised on my first tour—a thousand-mile trip that included myself (age seventeen), eleven other high school students, and two twenty-something leaders—when we accepted an invitation to stay at the house of a couple we had just met. The invitation had seemed like a miracle, since the weather, our strength, and our pre-planned campground had all conspired against us. We were out of options. Yet saying yes had also felt scandalous. It was contrary to the stranger danger warnings I had grown up with.

Skeptical and wide-eyed, I had followed my leaders. After a meal, we spread out on the couple's living room floor. I was overwhelmed by their generosity and trust. When they told us not to flush the toilet after peeing because they were on a septic system, I was convinced they were truly revolutionary. When they showed us their skydiving video, with a soundtrack by Tom Petty, I knew I had found the people I wanted to become. That night, I learned a traveler's best lesson: strangers are friends you don't yet know. The memory of that night is among the clearest and most influential of the month-long trip. Those people did not just welcome a band of cold, tired, dirty teenagers into their house, they invited us into their world. They taught me the value of saying yes.

On subsequent bike tours, my friends and I continued to learn from the magic of these meetings. Together we surrendered to chaos, embraced trips less planned, and welcomed the randomness of the world. Together we learned to trust in people, and to see the prevalent good rather than the occasional bad. That's not to say that we walked blindly into roadside offers. We also learned how to deflect people who seemed to carry red flags. We learned to trust our instincts and navigate outside our comfort zones.

Vicky earned my trust early in our conversation. My yes to her offer of tacos led her to usher me behind the counter, through a door, and into her kitchen. I sat as she sparked the oven to cook up a few eggs, heat some beans, and warm up a stack of tortillas. Customers came and went as Vicky's hospitality energized me.

As I filled up on tacos, I told her about that day's ride—the cows I had seen chewing on plastic bottles, the beautiful views, the long hill that didn't end, and the headwind that had kicked up and had been pushing against me like an advancing wall. Because my Spanish wasn't very good, I described the wind with my hands, using them to fake punch me in the face as I threw my entire body backward. I hoped my acting was enough to explain my haggard look.

"Quieres quedarte aquí en mi casa?" Her offer to let me stay at her house nearly made me cry. My grateful answer was, "Si." The next morning I would show my appreciation with a postcard-sized watercolor painting made on the road, a tradition I tried to keep with every stay. In the moment, however, I simply showed my appreciation with a wide grin.

Tacos, a bed, a shower, and a nice conversation were welcome miracles. I was shown to her son's old room, now vacant except for a bed, a few photos, and a chest of drawers weighed down by a large spread of baseball trophies. I plugged in my electronics, took a shower, and was asleep by 9 p.m.

Straightaway into the Heat

DAYS 12-18 / MARCH 23-29

MILES 575–1000

Ignoring the No Biking signs and passing a two-trailer semi chugging along, I raced downhill until the vegetation turned to desert spiky and the air to desert hot. Doubting my route choice, I merged onto a main highway. It was parallel lines of heavy traffic bent toward mountain ranges rendered blue by distance. Though it felt a bit like cheating, I had decided to speed up my ride by leaving the mountainous roads of the Sierra Madre and cruising the flat, paved highway running along its western slopes.

I justified taking a route I believed was out of the monarch's range by the ground I would make up. Sure enough, by 11 a.m. I had pedaled fifty miles, a distance matching the entirety of the previous day. The monarchs, tracked by citizen scientists, were already being sighted in Texas and Oklahoma, and I needed the easier days to log miles. The tailwind that urged me forward seemed to support my choice, and a sign, stretched across the highway like a rainbow of metal, completely validated it. I stopped to take in what I was seeing.

"Are you serious?" I asked, baffled.

The sign, straight from the heavens via the department of transportation, made a gratifying declaration: Ruta de la Mariposa Monarca. Apparently, I *was* on the Route of the Monarch Butterfly. Not long afterward, another sign amazed me even more. It asked drivers, the same drivers who ignored any inconvenient traffic law, to reduce their speed, from 90 kilometers per hour to 60 kilometers per hour, when monarchs were present.

I cynically knew that the request would be dismissed by most drivers. With the humor only solitude can create, I considered buying a bucket of paint and changing the sign to say, In the Presence of Monarchs, Drivers Should Buy Bikes and Cycle.

As fast as traffic was going, it was made bearable by the generous, well-paved shoulder. Unlike roads in the United States, in Mexico, highway shoulders are utilized by nearly everyone. After the shock wore off, I began to understand the system. The slowest traffic drove entirely on the shoulder, average-speed traffic drove halfway on it, and the fastest traffic could pass the other two easily by straddling the cleared center lane. At first alarmed, I found that to everyone's credit, the approach worked well. Best of all, cars at all speeds yielded the shoulder to me, and I felt safe as I pedaled.

In four days, I covered the 330-mile straightaway to the Mexico–Texas border by spending nearly all of my waking hours on my bike. Though the signs hinted at seeing monarchs, I saw no evidence of their passing. Unworried, I dedicated the time to logging miles. Up with the sun, done with the sun: it was a strategy to cover enough ground with my slow and steady pace, and to catch the cool relief at the bookends of each day. Far from the sea and mountains, I had found a world that was bathed in toxic heat as soon as the sun sprang into action each day. Though I carried several extra bottles, I would gulp water every time I found a store, and I began to feel like I was holding my breath between villages.

It was between two towns that I got my first flat tire of the trip. The *psssssssss* of escaping air taunted me as it blended into the heat. I had all the tools I needed to repair it, but as far as I could see, not one shade-producing tree poked into the blue sky. I didn't have enough water to lounge in the baking sun, so I got creative.

I steered my bike off the road and dipped into a narrow culvert. The cave of cement, cool and dark, was a relief. I suspected that I had found the closest thing to air conditioning in the surrounding area. I fixed the flat while cars zoomed overhead, their shadows casting moments of reprieve on the exposed grass. I was smug about my newfound oasis. Taking advantage of the potential in such a marginal space was very satisfying, a secret reward that only people at the mercy of such margins know.

I spent two of the next three nights in similar, yet bigger, culverts. Though such conduits need to be respected when storms brew, I saw no hint of past flooding, and concluded that they were abandoned, or nearly abandoned, passageways for cattle rather than water. Except for the trash thrown from cars, I saw no sign that humans ever visited them. The randomness of my location made me invisible, and for me, invisibility made for the safest camp spots.

In that stretch of Mexico, there was only one night when a feasible culvert didn't present itself. The sun was setting behind a fence that had stretched, uninterrupted, since lunch, and I found myself desperate enough to break one of my bike-touring rules. I spotted an unlocked gate, and stole through it.

Trespassing laws bother me. The idea that I can't trespass on private land but private landowners can trespass into my space, by dirtying my air and contaminating my water, has always left me doubting the validity of such restrictions. I admit, I find justice in trespassing. Still, for self-preservation, I don't cross fences or No Trespassing signs. Except when I do.

On the other side of the gate, a dirt road led away from the highway. I followed it until the noise of traffic was just a hum, then I left the dirt road as well. Behind a dense bush I set up my tent, obscured enough that no one would notice me unless they were looking. I liked the drama, and was overplaying it to make the evening more exciting. In Mexico, property rights are not religion as they are in the United States, and I honestly didn't think anyone would be upset about my intrusion.

By the time I finished my hodgepodge dinner, the half-moon's reflection was all that streaked the darkness. I sat in tired silence, until the sound of an approaching car on the dirt road startled me. I scrambled to sneak a peek, and watched it pass without incident. Out of sight, I listened. Gravel crunching. Silence. Door opening. Silence. Gate squeaking. Gravel dragging.

Gate closing.

Lock clicking.

I listened as the car accelerated into silence and I smiled. I was locked in. I had found the most secure camping spot ever.

The next morning, because I could barely lift my overweight bike up a step, let alone over a fence, I unloaded everything. I wrestled each bag and then my naked bike up and over. Finally, with a swift climb, I joined the heap. On the right side of the fence I put the bike-loading puzzle back together, merged back onto the highway, and joined the seemingly endless flow of unsuspecting traffic.

As time and miles blended, I slogged toward the endless horizon. I relied on National Public Radio (NPR) stories and Terry Gross interviews to occupy my attention until surprising splashes of color, the product of recent rains, refocused my mind. Primary colors grew like tributes to spring. Orange-barbed and leaf-colored caterpillars pulsed across the pavement, exposed. I was prompted to stop, photograph, and relocate each to its flowering camouflage. Flocks of white butterflies flowed like cottonwood seeds, east to west, for miles and miles. At each bridge, swallows dotted the blue of the sky and warming turtles dotted the blue of the water. In a land where the news found only fear, I found a passive beauty.

Late in the evening, eighteen days into my trip and 998 miles from my start, I entered a pack of rush-hour commuters. Two lanes became many, each filled with cars being herded through the gates of border control. Behind another cyclist, I wove to the front of the line, answered a minute of questions, and without ceremony crossed over the Rio Grande. There was no time for reflection about entering the second stage of my tri-country trip. One minute I was in Mexico. The next I wasn't.

The anticlimactic entry was fitting. If my monarch traveling companions saw no border between Mexico and the United States, why should I? For them, the Rio Grande was only a river. Ciudad Acuña, Coahuila, and Del Rio, Texas, were only towns. The monarchs floated from flower to flower, not country to country. The political distinction was merely proof of progress, a marker for the passage of time. There were miles behind and miles ahead. The border gave me permission to record one part of my trip as completed. Mexico north: check. I looked at the pavement unrolling before me, and the miles across the United States began to come into focus.

A Milkweed Greeting

MILES 1000-1105

If Mexico and its mountain spine are geologic shepherds guiding their flock of monarchs through a relatively narrow area, then Texas is the open pasture into which the flock is freed. When they hit Texas, the monarchs spread out with the wind and claim every road heading north as part of their path of migration. No longer racing toward spring, the next chapter for the females is to disperse and find milkweed plants on which to lay their eggs.

As the monarchs' range expanded on this migration, so too did my route options. Moving north, I wove my way into Texas on a route I had spent the previous year charting.

The bulk of my pre-trip route planning had come from studying the trends and patterns of the monarchs uncovered by Journey North, a migration tracking website. Since 1994, students, teachers, nature centers, citizens, and scientists across the monarchs' range have been submitting monarch sightings to Journey North, which then compiles the data on maps. By plotting the mountain of collected times and locations, the big picture is revealed. In general, the monarchs spread north though the Central Flyway, while slowly being pushed northeast by prevailing winds (some are pushed far enough east to generate a smaller Eastern Flyway). "In general" needs emphasis. By studying Journey North's maps, I was able to discern the infinite, subtle changes of each year's migration. I began to see the monarchs as one organism, growing, moving, and transforming over North America each season. I saw them as an artist painting a continent with their collective soul.

Arriving in Texas on March 30, and comparing this date to Journey North's data, I felt confident in my project. While some monarchs had arrived weeks earlier, the last arrivals were likely to linger well into April. I was in a solid position, neither at the front nor the back of the pack.

That breathing room was a relief regarding both time and place. Journey North's maps showed me that the spread north by the monarchs was like an unhurried wave breaking across the eastern half of the United States and southern Canada. While there was usually a concentrated corridor, the butterflies' range was large enough that as long as my timing was right, I stood a good chance of seeing monarchs regardless of my exact path. Biking north, I could join the millions picking their own way. Together, our comings and goings would weave a tapestry to cover their vast summer range.

Equally mitigating was the fickleness found in Journey North's maps. It seemed that some years, the migration stuck more to the Midwest; other years it leaned farther east. Some years the monarchs moved in a concentrated bundle like a lazy rocket; other years they dispersed like a slow-motion firecracker. Temperature, wind, and rainfall influenced many of these alterations, directly affecting the monarchs or the milkweed plants and the predators and pests that shaped their success or failure. My own success or failure was equally dependent on uncontrollable factors, so a good guess was the best I could do.

Since dispersing monarchs would likely be found whether I went fifty miles northeast or fifty miles northwest, I let the invitations I had received from people across Texas tailor my route, starting in the border town of Del Rio, Texas. There, I enjoyed my second rest day of the trip, spoke to Dr. David Forbes's college class, enjoyed the luxury of a complimentary hotel room, and celebrated my arrival with a meal. Before leaving, I swam in a local creek with turtle scientists (I saw a few turtles but wasn't fast enough to catch one). Then, back on the road, using a compass built by offers of kindness, I turned toward the Native American Seed farm.

Between invitations of hospitality, the freshness of Texas Hill Country captivated me in its own way. The caracaras cawed from their branched pedestals, the lupines drenched the hills in purple, and a scorpion danced among sun-shocked pebbles. Along the highway, a narrow spit of roadside ditch waved in the wind like a hand inviting me to stop, explore, and discover. Blurred by the pace of human velocity is a whole world crunching, crawling, wriggling, slithering, budding, branching, mating, living, dying, and migrating through a realm most of us look at but rarely see. I wanted to learn to see, to distinguish between the greens and purples and yellows. I scrutinized the passing landscape, looking in particular for one important plant. Like a female monarch, I wanted to find milkweed.

Milkweed—plants in the genus *Asclepias*—is the sole food source of the monarch caterpillar. There are more than 100 species of milkweed, seventy native to the United States, and monarchs will dine on many, though not all, of these species. Arriving in Texas each spring, the female monarchs scan their world for milkweed plants in which to lay their hundreds of eggs.

When I finally caught sight of an archetypal milkweed flower exploding from the roadside gravel, I abandoned my bike before coming to a complete stop. I waded into the ditch, to crawl among the flamboyant blossoms. Like eruptions of shooting stars, they were tethered to the ground by tendril-like stems. I watched, transfixed. A mostly southern species, antelope horn (*Asclepias asperula*)—with its narrow leaves folded delicately into empty spouts, and its elegant green flowers cradling accents of purple—was worth searching for.

Finding milkweed was a victory, as important to my trip as finding monarchs. Just as clouds are bound to the sea, the monarch is bound to milkweed. It's a knot that can't be untangled, an unbreakable bond forged by evolution.

Inspecting my first roadside milkweed with care, I searched for an egg, the product of a multi-month affair. Beginning in early spring, the choreographed dance of mating monarchs is performed in the sky. First a male will either attempt to capture a female in flight or pounce on her at rest. Once grounded, the male will wrestle the female, attempting to use his abdominal claspers to get ahold and copulate. Males, during this coupling phase, are often met with resistance from females, and it is likely that most attempts don't result in copulation. When a male does find success, he flies with the female hanging below. Silhouetted, the female looks like a reflection of her male escort, an upside-down, tandem paraglider. This conjoined configuration can last as long as sixteen hours, during which the male transfers his sperm to the female in a protein-rich spermatophore (a sort of packet filled with sperm).

As temperatures rise at the overwintering grounds, both males and females begin instinctually analyzing the cost and benefits of breeding. For males, the cost of mating is energy expenditure. The spermatophore can be up to 10 percent of a monarch's mass, and both fighting and flying with females uses lipid reserves. However, the benefit, though unlikely, could lead to the male passing on its DNA. While monarchs are particularly promiscuous, and females can store sperm from multiple males, mating in Mexico is likely less effective than breeding farther north. Most successful copulations are those performed close to the time of egg laying (in other words, in Texas, not Mexico).

While mating close to the time of egg laying leads to more success, some males find the benefits of breeding at the sanctuaries to outweigh the costs. These males are typically in the poorest condition. They often have diminished wing state and small abdomens (which mean less fat reserve). They are least likely to survive the migration to Texas. Males in better condition and more capable of surviving the migration will likely choose differently. For male monarchs, breeding in Mexico is a sort of Hail Mary.

Early studies (conducted by male scientists, it should be noted) assumed females were passive, so their behavior was not scrutinized. Now, scientists are less sure. It seems that females, too, have their own game theory. Females are seen resisting male mating attempts at different levels (and in different ways than observed in their summer range). At least in Mexico, the size of the males, with the smallest males being the most maneuverable and thus hardest from which to escape, may be a factor. Another factor could have to do with the receiving of the spermatophore.

Regardless of such equations, when monarchs arrive in Texas, both males and females limp with tattered and faded wings—worn-out warriors scarred from the battles of life. Using visual and chemical clues, the females home in on milkweed plants. Using their legs, which are covered in chemoreceptors, they drum on the milkweed's leaves to taste the plant's chemical composition, and, reading it like a nutritional label, understand its suitability. Sharp, spinelike tarsal claws on the tips of every female's brushfeet (the first pair of legs which, unlike the other two pairs, are typically tucked beneath the body and are difficult to see) cut the milkweed and trigger the release of more plant chemicals, much like cutting grass does. Drumming affords females extra-strong whiffs, to determine the best milkweed for their babies.

When satisfied, the female tips her body, curls her abdomen, and deposits, with a tiny bit of glue, a small egg on the underside of a milkweed leaf. Like tucking children into beds, monarchs choose the milkweed plants that will become the best possible nurseries and buffets. Though an average female will lay 300 to 500 eggs, she will typically lay only one egg per plant, to ensure enough food for each hatched caterpillar. For the overwintering females, this is their final act: to bestow the next generation a feast—sustenance that will feed the following leg of the migration relay—because even though the overwintering monarchs live longer than those born in the spring and summer, they can't live forever. By the time the females (and the strongest males that took advantage of late breeding) have arrived in the southern United States, they are old creatures. It is their turn to float down to Earth, return to the soil, and bide their time before blooming as flower petals. I was to follow many generations and to learn from them all.

I crouched on hands and knees to accommodate my curiosity. Each leaf appeared uneaten; no eggs in sight. Undeterred, I ripped a tiny edge of a single leaf, to watch its milky latex sap emerge. Testing the sap between my fingers I understood how it could gum shut the mouths of young caterpillars. With a deep breath, I tried to smell its bitter, heart-stopping poison (I couldn't). The gluey, toxic sap deters many herbivores, but not monarch caterpillars. On a neighboring milkweed I caught sight of a horseshoe-shaped hole in a leaf, a trench made by a caterpillar to drain the glue from its meal. In this way, the caterpillar is able to eat the milkweed and sequester the leaf's steroidal poison, cardenolides, in its body. The plant's toxins then become the caterpillar's own powerful defense.

A matter of evolution, the relationship of milkweed and monarch continued to play out in that roadside ditch, complicated yet perfect. I felt dizzy. In love. It was early in my journey, and already, I loved the monarchs because I loved the milkweed. I loved the milkweed because I loved the monarchs.

Satisfied with my search, I carried on, keeping my eyes glued to the shoulder of the road. Like a female monarch absorbed in her task, I hunted for milkweed and stopped often to inspect the shaded bellies of their leaves. Mile after mile, the orbs of milkweed flowers, sitting like eggs in leafy nests, stole my attention, until all the hunting paid off. At the first sight of those white, black, and yellow stripes on a caterpillar, I jumped off my bike like it was on fire and dove into the shoulder of the road. Chomping corn-on-the-cob style on a quickly disappearing leaf was my first monarch caterpillar of the trip.

Fifth instar, I figured.

Three to five days (depending on weather) after an egg is laid, a caterpillar hatches, nearly unrecognizable. This first size, or first instar, with its black head and translucent, cream-colored body, looks more like a wormy grain of rice than a monarch. It looks so unlike its larger counterparts that people will sometimes pluck one off their backyard milkweed, thinking it a pest. For the tiny caterpillar that survives, its life as an eating machine begins, starting with an appetizer of its chorion egg casing. Then it spins a pad of silk on which to perch (or possibly to dangle from, if disturbed), and the main feast begins.

As the caterpillars eat, they grow to the limits of their exoskeletons, and are forced to molt. From first instar to fifth instar, monarch caterpillars molt four times. During each transition, monarchs spin a new pad of silk and use the hooks on their rear prolegs (the fleshy pads on caterpillars that function like legs but are not true legs) to secure themselves to the silk. At this point, they stop eating for several hours while their outer cuticle skin separates from the new inner cuticle. As the old skin splits, a helmet-like head capsule falls away. Nearly liberated, the caterpillars must walk out of their old exoskeleton constraints. Watching the process, it is impossible not to root for the fresh-faced caterpillars contracting with effort. With great labor, they emerge. Triumphantly, they turn around and eat their own discarded skin. Nature wastes not.

With each molt, the monarchs get makeovers. Their front and back tentacles (antenna-like appendages) start as small bumps and stretch to elegant flairs. Their black, white, and yellow bands grow more pronounced and vibrant. Their six true legs see subtle changes in position, and their ten prolegs gain white-dot embellishments. They grow in size, too, but size alone cannot indicate age, as the condition of the milkweed also influences how much the caterpillars grow.

The fifth molt is perhaps the most dramatic. A larval finale, if you will, from fifth instar larva to pupa. The graduating caterpillar hangs from the silky pedestal of its silk button, and curls until it forms the letter J. The fifth molt begins as the exoskeleton splits to reveal the soft, neon-green of a young chrysalis. Its larval skin—an homage to its caterpillar life—bunches and falls away. Pulsing, the cremaster (the black post it dangles from) is driven into the silk so the chrysalis doesn't fall.

Most of us learn the basics of this transformation in grade school, but as adults, we stop truly looking, assuming there is nothing new to see. It is when I really look, with all my senses, at the feat of an egg becoming a butterfly, that I find God.

It made sense that I had first spotted, from my bike, a fifth instar caterpillar. Though size is not the best way to identify instar stages, the caterpillar I watched was big enough to catch my attention, around 1.5 inches long (fifth instars can range from .98 to 1.78 inches long). By the time fifth instars finish eating and form their chrysalis, the full-grown caterpillar could be as much as 2000 times bigger than when it emerged from its egg. I can't imagine the amount of ice cream I would need to eat to grow from

150 pounds to 300,000 pounds in just two weeks. We often overlook the grandness of small things.

As I continued to crawl through the ditch, a neighboring caterpillar stole my attention as it ate the flowers of a silently protesting milkweed. The clap of a car door closing pulled me out of my reverie. Unbeknownst to me, I was being rescued.

"Yeah?" I inquired, confused.

The Texas policeman stood by his car at the edge of the pavement. "Someone called 911," he hollered down at me. "Said a cyclist had crashed."

I stood to demonstrate my wellness. "I didn't crash," I explained. "I stopped to look at a caterpillar."

"A caterpillar?"

"A monarch caterpillar."

"A monarch caterpillar?" he repeated, moving back to his car. The caterpillar ate, unaware, as the car disappeared out of sight. I sat with the stillness. Someone had come to rescue me, but it was not me who needed rescuing.

For now, the monarchs and milkweed find refuge along roadsides, in backyard gardens, in feral and forgotten lands, but each year their options shrink. Habitat loss and the growing use of pesticides and herbicides are among the biggest causes of the monarchs' dramatic decline.

When a pickup truck rolled by fifteen minutes later, I was again pulled from my pondering. I assumed that this driver, too, was worried about me, so I waved, smiled, and shook out my legs. A second later, he made a U-turn, and parked right in front of me. Sporting a classic cowboy shirt, cowboy hat, and Texas accent, his looked at me suspiciously.

"People been trespassin' on my here land and stealin' my equipment." He paused as if waiting for my confession. Instead, I pointed to my bicycle in the ditch.

"First of all," I countered, exasperated, "this is not *your* land, this is the shoulder of the road. Second of all, I am on a *BIKE*." The bike part really needed emphasizing. "What could I possibly steal?"

No response granted, he drove away while I briefly considered trying to steal a tractor. It did not feel good to know someone looked at me and assumed I was a criminal. I wondered if that driver would have interrogated me even more had I not been a solo white woman. I wished good luck to the caterpillars, and aimed north. A spring migrant.

The monarchs and I were not the only migrants. Feathered, scaled, and hairy wings towed warmth and color from distant lands in their contrails. Yet migrants were not the only actors. Wildflowers—firewheels, winecups, and lupines—stood captivated by the sun and invited even the most timid pollinators to waltz among their petals. The hibernators, letting out long sighs as they stretched in the spring sun's rays, shook off the last grasp of winter and returned to work. I was surprised to find such drama along a Texas road.

Wanting to do more than pass through, I made the decision to wait five hours for the sun to set so I could catch the famous nightly show of a million bats erupting from a local cave, affectionately named Devil's Sinkhole. Even though waiting took time from making miles, I knew if I rushed, my memories would blur into forgetfulness. To wait would be to remember.

At dusk I watched the sun set and darken the Devil's Sinkhole. The world around me began to match the cave's innards just as the bats began to emerge. Like a corkscrew of wings they swirled upward. Bat wings buzzed. Barking frogs barked. A great horned owl patiently watched his dinner flood the night, while unsuspecting bugs awaited their own fates in the miles beyond. Thousands of bats swirled upward, a vortex that pulled the depth of their cave upward until I could smell the moist earth that collected each day in its hull. Unseen but not unnoticed.

Though the monarchs had long ago settled into trees to sleep away the night, I decided, after the bat show, to steal some nocturnal miles. Countercurrent to the stream of bats, I rode toward a veiled horizon. I turned my rear red blinking light on, just in case a car snuck up on me, but I kept my front light off, ready to flip it on when the distant headlights of an occasional car appeared. The nearly deserted road let me ride beneath undiluted stars. I could feel rather than see my progress, feel the hum of my wheels, the watchful eyes of the silhouetted trees, and the subtle changes in temperature as I rode the waves of pavement. I crossed invisible bridges over invisible streams. I felt like I was going faster than I was. I felt boundless. A show of lightning erupted miles away. I watched the flickering strengthen as the thunderheads flexed. We were running directly toward each other and there was bound to be a fight.

For two hours I rode, closing the gap, as the lighting went from show to threat. The frogs, sensing the storm's humid prologue, chirped noisily, and I yielded to the warnings by dragging myself and my bicycle off the road, into the frog's wild (though far enough from the river that the storm wouldn't wash me away). As I staked the last corner of my tent, heavy drops of rain tested the ground. I scuttled under the taut plastic, listening to the rain's drumbeat with a satisfaction that sprang from being tired, dry, and a day's ride from my next scheduled stop: a farm that was tilling native soil, planting native plants, and harvesting native seeds.

Finding
Refuge

DAYS 22-36 / APRIL 2-16

MILES 1105-1599

Upon my arrival at the Native American Seed farm, just outside of Junction, Texas, I was greeted by long rows of annuals and perennials I had previously seen organized only in the anarchical patterns of the wild. Bluebonnets and big bluestems lined up under blue skies. Scarlet sage complemented the scarlet soil. Firewheels and antelope horn milkweed waited in formation for their seeds to be harvested and planted by people wanting to foster natives and bring Texas back. Like a sergeant, I inspected each row—all with the potential to help rescue overgrazed and neglected shrubland that had been taken hostage just beyond the farm's forest edge.

Emily Neiman, second-generation native seed farmer, met me at the edge of the fields. My endeavor had come to her attention via an online newsletter, and she had emailed me an invitation. I had accepted, tweaked my route, and was thrilled to now find myself being led to the Neimans' guest cabin. With a wraparound porch shaded by leggy acacias, the cabin was both cool and sunny, open and cozy; the refrigerator was stocked with all the fresh, delicious treats I had been missing. I couldn't have imagined a better spot to catch up on emails, prepare for my presentations in town, do laundry, and recharge, all in the company of inspiring people. I thanked Emily, the network that had connected us, and the monarchs that led me there.

Around the barbeque and the sizzle of dinner cooking, I met Emily's parents, Jan and Bill Neiman. They had founded the seed farm in 1988. Prior to selling native seeds, Bill had worked as a nurseryman and landscaper, planting yard after yard with plants named for the distant countries from which they'd come: African Bermudagrass, Japanese boxwood, Pakistan crepe myrtle. Nothing was named Texas. Everything had to be irrigated, fertilized, and doused with pesticides. An observer of nature since boyhood, Bill began to question his practices. Why were the native plants of Texas that didn't need coddling being traded for such demanding replacements? He wanted to leave future Texans something he could be proud of. He adopted a code of land ethics, embraced native plants, and switched to organic farming; Native American Seed was born.

By going native, the farm team lets the plants tackle the hardest questions. Native plants, unlike non-natives, have adapted over millennia to survive the challenges of the land, without intervention. Texas natives can cope with the hot summers, the hungry herbivores, and the soil particular to Texas. They don't need pesticides, irrigation, or unnatural fertilizers. They work with nature, rather than against it. Their presence balances their ecosystem.

The monarchs understand this. Native plants are the glue that holds their migration together. While other farms spray away the milkweed, Native American Seed embraces each plant and their monarch suitors. I saw more monarchs on a farm tour with Bill than I had in the two previous days of riding. Bill welcomes each migrant, he doesn't mind when the caterpillars chow down on his rows of milkweed. He can share. The farm is the monarchs' present, and the seeds it produces can plant the monarchs' future.

Halfway through our farm tour, Bill stopped and dropped to the ground. On his belly, he gently prodded a small purple plant emerging along the trail—an Earth inhabitant he deemed worthy of examination. I don't recall the plant's name, but I do remember the grandeur of its tiny petals and Bill's curiosity. I remember admiring his relationship with every native plant, which gave him eyes to see a world most of us miss. He sees caterpillars as success, small plants as potential crops, and bugs as bird food. I knelt down, learning to see and celebrate the secrets cultivated by wildness.

Our tour continued to the seed company's office, where the walls were lined with posters of native flowers and shelves boasting organized packets of native seed. Native plants thrive on their own terms, and so the farm relies on incentivizing growth rather than controlling it. All seeds are either grown on the farm or scouted and conservancy harvested by contract on unplowed native lands. Once harvested, seeds are dried, cleaned, sorted, tested for quality, and distributed across Texas and surrounding ecoregions for both small- and large-scale plantings.

We also visited the wild buffer linking the farm to the Llano River, which runs clear and smooth, like a glass of cold water on a hot day. It traces the farm's edge as it heads toward bigger waters, and I could almost feel it linger as it tickled the still-native land—slowing, to remember all those forgotten. To remember the gift of desert gardens: strong, determined, protective, maternal.

The farm at Native American Seed was in direct contrast to my camp spot several days later, to the north, in the consuming sprawl radiating south from Dallas. Half lost at dusk, trying to avoid the main highway, I wandered through a partially built housing complex. I passed through the gates boldly, unimpressed by subdivision names like Prairie Villas and Meadow Oasis. No one was outside to stop me. No one was awake enough to see what we were losing: true prairie in exchange for a "prairie villa." We were trading a real prairie's treasury of life for a toxic monoculture of sterile green grass. As the houses closed in around me I wanted to yell, "If you hate the prairie so much, why did you move here?"

I found a small, untouched corner of wildness. Threatened on every side, my camp spot had no future. With the last of the light, I walked to a small pond that was coming alive with singing frogs. Tiny cricket frogs clicked anxiously at the edge of water and mud, while the larger green frogs squeaked before launching into the safety of their turbid water. I saw the approaching army of bulldozers looming nearby, and the lights of new homes polluting the once-raw darkness. The male frogs called, attracting mates to lay eggs and begin another generation. Their calls seemed to be protests, an act of defiance. Even as the walls of development squeezed the life out of Texas, that life called to its doomed future, reminding anyone who would listen that it was their home, too. I listened. With a heavy heart, I wondered how anyone could call such destruction progress.

The frogs brought into focus a growing sense of dread I was feeling as I cycled with the monarchs and saw the world through their eyes. What I saw scared me. I saw huge swaths of the monarchs' home and migration path freshly plowed, freshly bulldozed, and freshly paved. Freshly destroyed. This attack on nature was an attack on me, and I felt the pain of its punches. I didn't realize it then, but my trip was not about conquering miles. My trip was about absorbing the destruction of such important lands and using my voice to fight for the future. I would be like the frogs singing in the pond, singing not because they could secure their future, but because all they could do was try.

I wonder if that pond still sings with frogs today?

I thought about them as I packed up in the quiet morning and headed deeper into the sprawl, until I was swallowed up by the bustle of Dallas. I wove through downtown, imagining that as I stared up, a monarch was staring down, and together we were both trying to understand the sky-rocketing buildings. I found and entered just such a building, crossed the lobby with my bike, and stepped into the elevator.

I stayed with Elizabeth and Stan Hart in their high-rise Dallas apartment for several nights. Their walls brimmed with art and the bike room overflowed into the hallway. I wondered if the monarchs found gardens to be as much of a relief as I found the small apartment to be.

For several days, Elizabeth and Stan gave me the grand tour of the city. Pizza, gardens, bike routes, art festivals, and most rewarding, the Mega March for immigration reform and racial equality. We joined immigrants from around the world (I am a third-generation immigrant), and let our bodies and voices grow the collective, just as the monarchs do. From farm to city, from here to there, from lost to found, the monarchs combine forces and march with wings over fences of their borderless world. They are a symbol of a shared continent. They are North American. We are all creatures of one world.

My next stop was at the home of Linda Lavender and Mike Cochran in the Dallas suburb of Denton, Texas. They had contacted me several months earlier and offered me the use of their home while they were traveling abroad. I found their yard satisfyingly devoid of manicured grass and was instead welcomed by a crowd of Texas bluebonnets. In anticipation of my visit, Linda and Mike had organized my Denton school presentations, arranged for their daughter to come by and unlock the house, and for their friends to bring over dinner. Moments after my arrival, just as planned, their daughter pulled into the driveway. She ushered me in, gave me the tour, and handed off the keys. It seemed almost impossible that in 2017, people would trust a near-stranger, but there I was, proof jingling on the keychain in my hand.

From Denton, I headed north to the protected lands of Hagerman National Wildlife Refuge, the first of many national refuges I was to visit with the monarchs. Dotting the United States, the National Wildlife Refuge System might be slightly less glamorous than the National Parks, but their unpretentiousness allows for a more intimate connection to nature in your neck of the woods. Through the system's subtle management of these public lands—tweaking water levels, regulating grazing, adjusting visitor access, providing nest boxes—plants and animals we might never learn to identify are given a chance to thrive. For migratory species, such as waterfowl and monarchs, the refuges provide critical stopover sites at which to feed and rest. They are islands of habitat in a sea of human sprawl, life rafts in the storm of progress.

Arriving early at Hagerman, I was told by the staff that I'd be treated to a night in one of the refuge's spare trailers. When the gates closed at dusk and frogs began to fill the space between the stars with their songs, I was alone with the refuge's wild tenants, excited to explore our shared backyard of protected public land. Like a game of Marco Polo, I followed the sound of the frogs through the darkness, staggering around clusters of cedars and blackberries. Each creature I met was protected by the refuge's border, an invisible defense against demolition, and I experienced no dread like I had at the pond just south of Dallas. A refuge of land; a refuge for my heart.

At the water's edge, I took off my shoes and waded in alongside a nondescript water snake. The frogs sensed my movement and quieted their courting melodies, as ripples reached from shore to shore. I stood still, trying not to intrude. After several moments the bravest grey tree frog broke the silence with a trill. Under his leadership, the others joined. There was the clack of cricket frogs, the squeak of leopard frogs; I was blanketed in the ballad of spring. I explored the pond with childlike abandon. Though none of the species were new to me, those exact frogs were, and each sighting filled my heart.

My love of frogs started when I was twelve. The kids I babysat received two green tree frog pets. I loved the way the frogs' eyes looked at the world, how their toe pads stuck to the glass. Soon I had my own tree frog set-up and a job at the local pet shop. Watching them catch flies or revel in a humid mist never got old. Watching eggs develop into tadpoles, and tadpoles into frogs was always a miracle. Those frogs were my teachers, and they taught me to love the brilliance of tiny creatures.

In college, I studied wildlife biology and each summer I got paid to investigate critters in beautiful places, nearly always on public land. I studied sea turtles in Hawaii, amphibians and reptiles in Wyoming, and toads in California. After graduating, I continued with seasonal work, studying frogs in Montana, fish in Utah, and tortoises in Nevada. Each job let me experience the incredible breadth and importance of our public lands and gain new perspectives on the world. I trained myself to see the land from the eyes of frogs, snakes, and turtles. A decaying log, a bend in the creek, a solitary milkweed plant, an acre marked for future development—seeing was a blessing and a curse.

Seasonal jobs offered other benefits. The jobs exposed me to adventurers and boundary pushers. As coworkers, roommates, and friends, we nurtured our unconventional sides, embraced the challenges of remoteness, and nudged each other to venture further and further from the status quo. We romped through the wilderness, gaining the confidence to push into the unknown.

Because of this background, my leap from studying amphibians to biking with butterflies seemed like an obvious one. My love of amphibians and reptiles (herpetofauna) translated easily into a love of monarchs. Both are transformational underdogs, sensitive to environmental changes, that can easily go unnoticed if you forget to look. Both metamorphose under impossible odds. Both leave me in awe.

Following frogs by bike would have had its limitations. Frogs rarely travel far from their natal lands, and never across continents. They rarely thrive among humans and their existential solutions have higher hurdles to clear.

The migrating monarchs, however, were a perfect storm of possibility. As a whole, the migration advances an average of twenty-five to thirty miles a day. Though individual monarchs can go much farther, with one monarch having been recorded flying 265 miles in a day, a cyclist can cover similar distances. Spreading out in the millions across a landscape traced with roads, there were few route-planning limitations. At home in backyards, school gardens, parks, roadside ditches, and the wildest places, monarchs, like clouds, are democratic in their reach. Following them by bike seemed meant to be.

Monarchs, too, are easy to spot and identify; their beauty easy to appreciate. I saw them as gateway bugs and ambassadors of nature—teachers, inviting people to look closer. Robustly studied, yet the subject of many unanswered questions, they showcased the power of science. Threatened with extinction, they were the motivation I needed to use my body, time, voice, and bike to contribute to the conservation movement.

Chasing Spring

MILES 1599–1783

"Welcome to Oklahoma," I declared as my tent filled with water.

From Hagerman, I had traveled to Tishomingo, a wildlife refuge on the Oklahoma–Texas border. I was camped in a grove of trees, and even the frogs had tucked themselves away as clouds dumped a storm of raindrop confetti.

The rain had started sometime in the night, slowly saturating the carpet of needles under my tent. As the sun stretched into its new day, the rain pledged a storm. My tent became a leaking boat from which I was trying to navigate the storm.

I looked at my watch; 5:15 was too early to get up, so I pawed at the nearby needles, created a small moat to distract the water, and bought myself some time.

By 7 a.m., the rising waters left little room for negotiation. I conceded to the rain by stuffing my gear into my bags, unzipping my tent, and stepping out into four inches of water. The shock of the cold was liberating. Feet wet, I had no need to tiptoe. I splished and splashed, laughing at my self-made spectacle. *I signed up for this?!*

With bags loaded on my bike, I didn't bother to roll up my tent. Instead I plopped the soggy mess on the back rack, tied it down, and took off. The novelty of the watery eviction outweighed any regret I might have had about choosing that camping spot.

On the road, moisture snuck through my raingear, striking from above and below as it first fell from the sky and was then spat back up by my churning tires. Lightning flashed and thunder rumbled. As long as I didn't stop, the warmth generated by my exertion kept the storm's cold at bay.

Soaked through, I reached a town. With no pressing plans, I found an escape from the brunt of the storm by asking a woman under an umbrella where the library was. On bike tours, libraries are like stepping-stones. They provide the tables, electrical outlets, internet connections, space, and atmosphere to do what you need to do without expectation. For the five hours I huddled between the shelves, busying myself with my to-do list, there was no judgment. I was likely not the only wayward soul using the library for refuge. When I left, it was only because the heaviest clouds had parted and the drying blue sky dared me to continue.

I rode through the puddles that lingered on the pavement. My wheels disrupted their own reflections. The wind felt fresh on my soggy skin, and my raingear, strapped to my bike, shimmied itself dry. Among the checkerboard farm roads, I rode out the rest of the day. My campsite that night was behind a church. As small as a shed, it didn't conceal me, but there was no need for hiding. In the rays of the setting sun I set up my tent to dry and ate dry Cheerios by the handful for dinner.

Like a key turning a lock, Oklahoma's spring rains opened the door to restless reptiles. Wandering box turtles cleared spider webs and dusty tunnels to meet the sun. The silver dollar–sized painted turtles scrambled hungrily toward warming waters. A gopher snake, which I caught and moved off the road, traveled with me in my thoughts. Did she make it to wherever she was going?

The monarchs traveled with my thoughts as well. Each sighting was a wonder. By Oklahoma, I had only seen forty-five adult monarchs, but that was enough to keep me motivated. One was better than none. Like a salutation, I stopped to record in a tiny notebook each monarch I saw. I scribbled location data along with notes on the time, the wind, and what each monarch was up to. I wondered if the people in the passing cars knew they were passing monarchs. Did they imagine that I was trying to immortalize each one with pen and paper?

Noting each monarch helped me consider their many details. While I was still seeing the overwintering generation of monarchs, some riding on wings that were more sky than orange, I was beginning to find the bright, fresh color of the next generation. The older, tattered monarchs gave testimony to resilience, while the younger butterflies reminded me that the migration was built by a surrender to lineage. The monarch migration is a relay. This new generation would fly the next leg to Canada. In total, it would take three to five generations to complete the round-trip migration. Only by looking deep into its DNA can a single monarch know the entire journey.

As cars whooshed past, I crouched by a roadside milkweed and contemplated an egg on the underside of one tiny, fresh leaf. Like a period containing an entire story, it didn't seem possible. How had a monarch found such a small plant? How could that little speck go from egg to caterpillar to pupa to adult, fly north, lay eggs, and die? How could its offspring do the same? And how could its grandchildren or great-grandchildren know to fly back to Mexico in the fall?

If I could have articulated that feeling of being witness to such an awe-inspiring process, then I would have had a better answer to the question often asked during my journey, "Why should we save the monarchs?" It was a troublesome question to me, one that made humanity seem doomed each time I heard it. To my mind, the answer was so obvious that words couldn't answer it.

We must save the monarchs because the monarchs *are*.

Dr. Lincoln Brower understood what I meant by this. As a leading monarch biologist, Brower published 167 peer-reviewed journal articles, helped create the Monarch Butterfly Biosphere Reserve in Mexico, was a board member for both the Monarch Butterfly Fund and Journey North, and gave many presentations to increase awareness. Sadly, Dr. Brower died in 2018. Happily, his legacy lives on.

In a video made by Dorothy Fadiman, Lincoln Brower looks at the camera and says, "On occasion I will be asked in a public lecture, 'Well, Professor Brower, tell me, what good is the monarch butterfly?'" In the video, he laughs before continuing, "Needless to say, I am extremely irritated when anybody asks that question."

When I watch the video, I nod understandingly at the white-haired man on the screen. He continues, "I've seen the Mona Lisa in Paris. . . . What good is the Mona Lisa? Really, it's just a painting on a piece of paper, but we revere it as part of our culture and part of our tradition. . . . I think we need to realize that biological treasures, such as the monarch, are just as valuable as the Mona Lisa. . . . These things are fascinating in their own right, and valuable in their own right."

When he tells the story of one of his graduate students responding to a similar question by asking, "What good are you?" I laugh, wiping away tears. It is always nice to know you are not alone.

Lincoln Brower went a step further. He knew that the monarchs, as a species, would likely survive, even if their migration does not. After all, monarchs are native to both North and South America, and in more recent times have spread to Hawaii, the South Pacific, Australia, New Zealand, and the Mediterranean. How they crossed the Pacific Ocean remains a mystery (though humans likely had a hand in it). However they arrived, it now means that monarchs as a *species* are not in danger of extinction. Thus, Brower advocated for the protection of biological phenomena. He knew the migration needed to be saved for its own intrinsic beauty. This led him to sign the 2014 petition asking the United States Fish and Wildlife Service (USFWS) to list the monarch as a threatened species. After several extensions, the USFWS had not released its final decision by the time this book when to print.

I know that even if the monarch migration in North America disappears, I could still go to, say, Australia, and see a nonmigratory monarch. Though always beautiful, I would feel a loss. The magic of walking through a cloud of overwintering monarchs in Mexico, with their wings and shadows mingling, would be missing. So would the wonder of watching a monarch lay an egg in Canada, knowing that egg might well metamorphose and fly to a tree thousands of miles away. There is a difference between a species that exists, and a species that carries out a *multigenerational, multinational migration*. Whether we can save this great natural wonder remains to be seen.

I agree that listing monarchs as threatened would protect crucial habitat, help people grasp the butterflies' urgent circumstance, help establish a recovery plan, and promote large-scale conservation efforts. I also agree that such a measure would need to ensure that allowances are made so that people can continue to rear caterpillars and learn from wild creatures held temporarily. Under Appendix B of the petition, individuals, households, and educational entities would be allowed to rear no more than ten caterpillars each year, which would mean huge changes to how people connect with monarchs and how the heart of the conservation movement beats.

I see the pros. I see the cons. I see that something must be done.

On Oklahoma State Highway 7, scissor-tailed flycatchers draped the sky and I let them distract me. A field covered in salmon-, fuchsia-, and crimson-colored Indian paintbrush faded into a colorful haze. Linking several local roads together, I arrived at Sandy Schwinn's house south of Tulsa, Oklahoma, minutes before another spring storm tried to wash the city away.

Sandy is one of twenty-four Monarch Watch conservation specialists spread across North America—spokesmen and delegates who help bring attention to the monarchs' plight. She had learned of my project and extended an invitation to stay with her when my route came close to Tulsa. From the dry comfort of her living room, I watched the sky split open as she tended to her caterpillars.

Like a mother to the monarchs, Sandy has devoted much of her energy to her monarch family. Though her garden, which spreads across her backyard as shared space with nature, is the focal point of her dedication, it is by no means the full extent. During my visit, her kitchen was hosting dozens of hungry caterpillars, collected off her backyard plants. They ate milkweed in the safety of plastic containers, cleaned and replenished twice daily. On the counters, stacks of monarch brochures and fliers sat ready for her next event—tabling at fairs, giving presentations to interested organizations, going anywhere interest was shown. On her porch, milkweed seedlings sprouted out of tiny black plastic pots. She had planted the delicate milkweed saplings as seeds. They waited patiently for Sandy to share them with anyone who wanted to grow a monarch garden. In this way, she did the job of the wind, dispersing milkweed so that it could put down roots in every corner of the city. She was regrowing the prairie one small pot at a time, growing a resistance and a revolution one garden at a time.

Sandy had adopted the monarchs thirty-six years earlier, and since then, she had made several important observations. One was that Oklahoma, once thought unimportant during the spring months of the migration, was in fact an essential stop, hosting many monarchs.

Another observation, which she had just made when I visited, was that the monarchs were arriving sooner than usual that year.

Sandy was not the only person to notice the early arrival along the migration route, and many gardeners, acting as the eyes of the migration, were troubled by the inconsistency. I, a monarch novice, had assumed incorrectly that going north quickly would be like a head start, and thus a good thing. But as I talked with Sandy, as well as other gardeners and scientists, I began to understand the harm that came with the strong winds pushing the monarchs north far earlier than normal.

In places north of Oklahoma, monarchs were arriving before much of the milkweed had emerged from winter dormancy. Females were being forced to lay dozens of eggs on single stalks of tiny milkweed plants. There was no way a single milkweed could support so many gorging caterpillars. The cold northern temperatures were not only causing egg dumping, they were also slowing down the progress of metamorphosis. Instead of going from egg to sexually mature adult butterfly (the females need an average of five days to become egg laying, another factor that is temperature dependent) in the normal thirty-five-day window, the colder temperatures were extending this egg-to-egg interval to as many as forty-five days. The butterflies were not just losing their head start, they were potentially being robbed of an entire generation and the population bump each generation provides.

Monarch researcher Dr. Orley "Chip" Taylor has been reviewing the effects of seasonal weather variations on monarch populations, and has been developing a model to predict the overwintering population based on a combination of environmental factors. He has found that the pattern most conducive to population growth begins when monarchs from Mexico lay their eggs in Texas (rather than farther north as Sandy and I were seeing). Laying eggs in Texas jumpstarts the entire breeding season, allowing more generations to be born during the reproductive season. This happens because females that lay eggs in Texas don't waste their time traveling farther north than necessary, and because eggs laid in Texas will likely find warmer temperatures. Warmer temperatures mean a faster development and shorter generation times. The more generations that can be crammed into the summer breeding season, the bigger the population can grow.

Time, temperature, winds, milkweed . . . so many variables. It's a performance in which every element gives and takes cues, each a part of the whole. As we shift the balance and alter the equation, Earth spins on, and we can only begin to understand the implications.

"How many species of milkweed do you have?" I asked Sandy, as we discussed the migration on a meandering stroll through her garden.

Amused by the question, she started to count. "Nineteen," she declared before remembering the whorled milkweed in one corner and the clasping milkweed in another. "More than twenty," she decided. I was gaining confidence in my milkweed identification, but on the roadsides of Oklahoma, there were only two, antelope horn milkweed (*Asclepias asperula*) and green milkweed (*Asclepias viridis*), that I felt I could recognize. Even those proved confusing. The common names—antelope horn, green, and green antelope horn—were inconsistently used and often interchanged. It was more practical to use their scientific names, which to me made them sound like superheroes (just imagine Asclepias storming the castle while Viridis and Asperula shoot laser beams from their eyes).

Whatever names we use, there are more than seventy species of milkweeds native to the United States, and thirty of them are used as host plants by monarchs with some regularity. They are waiting for an invitation, and Sandy has made her home and outdoor spaces just that. Her gardens have become a conglomerate of ecotypes and her yard, a novel ecosystem.

Only when we give nature space can we give it life.

Remembering Tallgrass

MILES 1783-2394

From Sandy's, I pedaled six miles through an urban labyrinth to another monarch way station, another refuge, for the butterflies and for myself. As soon as Amy Lucas Whitaker ushered me into her garage to park my bike, I felt at home. Amy had that wonderful "mom" energy. My arrival was celebrated, and she held my stories like prizes. I paused during the tour of her house to notice a large painting hanging over the piano. A mama elephant hooked her trunk around her baby. It had been painted by Amy with brushstrokes of every color.

Featuring a pool and green grass, Amy's backyard at first didn't seem like a place where wildlife would flourish. Then my eyes fell on the patch of milkweed sparkling like a gem, a retreat of wild green celebrating the spring day. The milkweed's sturdy leaves extended upward and outward, a potential nursery for future monarchs. I moved to inspect my favorite species of milkweed, common milkweed (*Asclepias syriaca*). Though it was a small garden, Amy told me forty caterpillars had already been cradled in its leafy arms. Even if just one survived, such a contribution to subsequent generations could mean hundreds more monarchs down the line.

Like a stone dropped in a sea of indifference, each planted milkweed causes a ripple. Each milkweed tended, each monarch lover I was encountering, each mile I was logging was somehow making a difference.

Together, it felt like our efforts could launch waves across strip malls and fields of corn, across continents and oceans.

Call it the butterfly effect.

Each year, the size of the monarch population is estimated not in the butterflies' spread-out breeding range but in their concentrated winter range. Though it is easier to count monarchs when the land they occupy is 100 square miles instead of 1.6 million square miles, it is still not easy. After all, an average-sized oyamel branch has been estimated to hold nearly 6000 clustered monarchs; the collective weight is enough to potentially break branches. Scientists instead measure the area occupied by monarchs each winter and rely on yearly comparisons to understand population trends.

To measure the winter forest, scientists begin by visiting all known historical sites of monarch clustering, both in and out of the Monarch Butterfly Biosphere Reserve. Then, in December, when the butterflies have settled into their colonies, counting begins. The participants in these surveys have varied over the years, but currently the team consists of scientists from the World Wildlife Fund (WWF), the National Commission on Protected Areas Mexico (CONANP), the Secretariat of Environment and Natural Resources Mexico (SEMARNAT). Going from outer tree to outer tree, the scientists measure the perimeter of the colony, and a polygon of monarchs emerges. The area inside is added to the total area of overwintering monarchs. In theory, the more area there is, the more monarchs there are.

Best estimates suggest that each hectare (one hectare is 2.47 acres or nearly two football fields) contains an average of 21.1 million monarchs. Such variation in estimates reflects the fact that the goal of these counts is not to know the exact number in the population, but to create an index that can be compared over many years. If the area is calculated with the same protocols annually, and if, on average, the area stays the same, then the population is stable. Unfortunately, such stability has not been the case. Looking at the Monarch Watch graph depicting the area of monarchs occupied in Mexico each winter, it is easy to see a disturbing downward trend. The population graph jumps up and down, and the trend line paints a disturbing picture. In the winter of 1996–97, the monarchs occupied 52 acres (20.97 hectares). By 2013–14, that number had plummeted to 1.65 acres (0.67 hectares).

Luckily, since 2014, there has been some recovery, but the trend line remains to be seen. When I left on my bike tour, in the winter of 2016–2017, the population was measured at 7.2 acres (2.9 hectares). The 2018–19 count was the highest in ten years, at 14.9 acres (6.05 hectares). The rebound was promising, but promise does not mean recovery. In 2019–20, the population fell back to 7.0 acres (2.83 hectares).

All wild, healthy populations rise and fall from year to year, as conditions vary. The concern for the monarchs comes from the long-term downward trend. Some scientists estimate that in the last twenty years, nearly 90 percent of the monarchs have gone missing from our skies. This plummeting population leaves little allowance for the toll that could be taken by a freak weather event, disease, or a consequence of climate change. Looking at historical averages, scientists have identified a population target of 15 acres (6 hectares). Anything less and the chance that the population will reach zero looms likely.

Zero marks extinction. Not of the monarch species, with its global distribution, but of the migration. We are facing the extinction of a phenomenon.

Scientists have linked this alarming decline in large part to habitat loss. Monarch Watch, the University of Kansas's education, conservation, and research program, estimates that *each day*, 6000 acres of monarch breeding habitat in the United States are converted to something else: housing or commercial developments, farms, roads, and other human uses. Even farms, which once invited milkweed to thrive between crops and along farm edges, are changing tactics and destroying milkweed. The presence of milkweed in agricultural fields (between crops and on field edges) declined 97 percent from 1999 to 2009 in Iowa, and 94 percent in Illinois. Each year, the migrating monarchs have fewer places to feed on nectar and lay their eggs. They are losing their habitat, losing their homes. Eviction, extinction.

As I biked, I imagined what that would be like. I could find a grocery store most days, but what if I had to bike a week, or a month, between each shop? What if every grocery store closed down?

Moving north into Kansas and the heart of the Midwest, I saw what scientists were documenting. I found that the fertile prairie had been sacrificed to wheat and corn. Our breadbasket was brimming, at the expense of the monarch. It was only when my path crossed that of wild creatures that the prairie ghosts would subside. The presence of hawks stationed on their posts, the surprise of garter snakes soaking up the sun, the sight of buckeye butterflies' predator-deterring "eyes" on their wings—all reminded me that there was life still holding on, waiting for the storm of demolition to pass.

The seemingly doomed American prairie is one of the monarchs' many crux points. This is an important clarification; too often the burden of the monarchs' decline is placed solely on the destruction of Mexico's oyamel fir forests.

Yes, the forest in Mexico is under threat. Currently the Monarch Butterfly Biosphere Reserve is divided into two zones. The core zone is the most protected, and theoretically all extractive activities are prohibited there. In the buffer zone that surrounds the core, controlled logging and active management are allowed, as well as the collection of firewood and mushrooms. Analyzing aerial photographs to track disturbances in both areas, it seems that from 2001 to 2012, an estimated 2057 hectares (5084 acres) of the core zone was affected by illegal logging. This includes 1254 hectares (3099 acres) that were deforested to the point that less than 10 percent of the canopy forest remained. Luckily, stricter enforcement and economic incentives have helped lower the rate of large-scale illegal logging. Now, it seems, small-scale logging needs to be the focus of concern, as it grows with the local population. In 2010, 27,000 people lived in the buffer zone of the MBBR, and more than one million people lived in the area surrounding the reserve. The growing population, with limited economic resources, will continue to rely heavily on the forest and surrounding land to sustain them.

As crucial as the overwintering forest is, the prairie is just as important. Blame cannot rest entirely on Mexico's shoulders. The United States and Canada are also complacent. I, a Kansan, am complacent. In my home state, once a galaxy of grass, the tallgrass prairie now barely clings to its soil. Once present from Canada through Texas, now just 1 percent of the historic tallgrass prairie remains, making it one of the most rare and endangered ecosystems in the world. As dramatic as mountains, as vast as a sea.

While I am a perpetrator, I am also a victim. I inherited the shadow of something spectacular, and am left with only my imagination to envision the prairie's once-endless splendor. I can only dream of the millions of bison that once chomped, wandered, and produced the prairie under the gaze of visiting monarchs. Looking out at the broken scraps of what once was, my heart is broken, too.

Yet as the prairie continues to disappear, there are pockets of preservation. People are finding space for the prairie to breathe. In Wichita, Kansas, the Great Plains Nature Center cradles a reminder of what Kansas once was. Near Pleasanton, Kansas, Marais Des Cygnes National Wildlife Refuge keeps yesterday's landscapes more than a memory. In backyard sanctuaries, such as the one my Aunt Patty and Uncle Gary have created in Overland Park, Kansas, our planet's heritage blooms and the monarchs celebrate. To save the monarchs, we must follow the lead of the rangers, refuge managers, and my aunt and uncle. We must give space back to the prairie.

I continued to the edge of Kansas until I was at my childhood home, a stone's throw from the border city of Kansas City, Missouri. My parents' excitement at my arrival had been mounting in the weeks prior, their love evident. My mom had planned every meal (starting with her vegetarian specialty: spaghetti-squash casserole) and stockpiled York Peppermint Patties and chocolate ice cream. My dad had been anxiously checking my nightly camping spots, plotting each on a map, adding my mileage into a spreadsheet, and using Google Maps' Street View to follow my journey, as only a hard-core fan could. I, too, looked forward to the visit.

I assumed part of my parents' excitement was the relief they must have felt knowing that, for at least a few nights, I was safe. "What do your parents think?" was a popular question as I traveled. The answer was complicated. I knew my folks worried, and I knew they had a right to be worried. I met many parents on the road and a good portion of them said something to the effect of, "Your poor parents," or "Thank God I'm not your mom, I couldn't handle it." My response was to explain the concept of risk, to remind people that safety could never be guaranteed, regardless of our choices. To choose the illusion of safety over following your dreams was a different kind of recklessness. "Besides," I would continue, if I sensed they were not convinced, "I have proven that I can do this. I am not blind to danger. Instead, I put risk in perspective. I think of bike touring as a precautionary measure against a different threat: heart disease."

I don't know if my parents think my adventures are worth the risk. I do know, as I continue to explore the world, that they keep their worries mostly to themselves. They have accepted who I am, and I am grateful that they don't burden me with their worries. I, in exchange for their sacrifice, do what I can to stay safe. I take calculated risks, just as everyone does when they drive a car, walk through a city, eat a hamburger, or wake up each morning and go about their lives.

I rang my bike bell as I pulled into the familiar driveway. The garage door opened when I entered the code, just as it had done hundreds of times before. My parents were at the door by the time I parked my bike. We exchanged hugs and stories as I sunk into the comfy couch and tried, unsuccessfully, to entice Bernina (the cat my brother brought home) and Pistachio (the cat I brought home) to jump into my lap. "How is the milkweed?" I asked, jumping up and rushing toward the far corner of the lawn to check on the first milkweed I had ever planted.

"Just don't dig it all up," my mom had told me, nearly a year earlier, as I had grabbed a shovel during a spring visit.

I had found a nursery selling native plant starts, all of which were neonicotinoid free. Neonics, as they are known, are neurotoxic insecticides applied to seeds and roots of many nursery plants and crops. These applied neonics dissolve in water and the plants absorb them into their leaves, nectar, pollen, and fruit. The entire plant becomes a poisonous vector. The toxins persist for years in the plants, making their damage far reaching. Bees that pollinate neonic crops are killed by the nectar. Birds that eat neonic-coated or contaminated seeds in affected fields become collateral damage. Humans, too, are exposed to these neurotoxins. Cherries, for example, that were sampled from 1999 to 2015 were found to contain neonic residue 45.9 percent of the time. Monarch Joint Venture, a partnership protecting monarchs, recommends (if disposal is not an option) cutting off the flowers of treated plants for several years, so as not to attract and poison pollinators. They also suggest tenting plants to prevent insects from accessing the leaves. I told the nursery how happy I was that they had pesticide-free plants. Nurseries can't hear this enough.

I had carried my twelve native starts, including three milkweeds, home in my bike trailer. Then, with a shovel and my mom's blessing, I dug twelve holes. Loosening the roots from their potted confinement, I had gently placed them in their new home. I toasted the occasion with water from the hose, and declared native planting easy enough.

Returning now, I saw the good news blooming. The Joe Pye weed, Missouri evening creeper, and goldenrod were all alive, sniffing the air as if worried about hungry bison. The milkweed, however, was very, very dead. D-E-A-D. The demise of the milkweed gave me a perfect lesson to bring to my upcoming school presentation. By the time I'd reached Kansas, I had made presentations to a handful of schools and nature centers about my adventure and the monarchs. Giving voice to conservation fed the glimmer of hope I carried in my heart.

"Dead . . . dead . . . dead," I told the kids dramatically at my next presentation. "Seeing my dead milkweed, I felt like a failure. All I wanted to do was this . . ." With a dramatic pause, I took a deep, calming breath—then exhaled an eruption of over-acted frustration. Arms flailing, I whined, "This is too hard. Milkweed is stupid. I hate gardening." The kids giggled, understanding.

"But," I continued, lifting an eyebrow and tapping my chin (my actor-y thinking face), "instead, I asked myself, *What went wrong?* Then I did some research. I learned that milkweeds like lots of sun. The corner where I had planted them was probably too shady. I picked a sunnier spot to try again."

I asked the kids to show me with thumbs up or thumbs down whether they thought my second try with the milkweeds would go better. They confirmed what we all knew. "Better!" they shouted, a sea of tiny thumbs signaling the power of trying, then trying again. (At the time, I didn't know that I would need a third *and* a fourth try.)

The idea to connect students to my adventures began while planning a bike tour to every state but Hawaii in 2010, coined "bike49." As my four friends and I plotted our 15,000-mile route, I knew that to stay engaged on a yearlong bike tour, we would need a way to give back and have a focus beyond cycling. I proposed the idea of giving presentations to students. Though none of us were teachers or especially eloquent, we dove in.

Our first presentation was horrible—and the most important one we made. We learned more than we taught, and I am forever grateful to the teacher who let us flounder. We were so committed to getting better that we embraced every mistake. To this day, I do my best to avoid describing things as "cool." Cool explained nothing, the teacher had patiently told us afterward. Cool was a blob. Words like "diverse," "exciting," "unknown," "unique," "colorful," and "strange" painted pictures.

On bike49, we presented to over 100 schools, and each and every one was a learning process. Months after landing on a presentation we were proud of, and after presenting to a roomful of kindergarteners in New Jersey, the teacher followed up with a question. She held up my air mattress and asked the class, "Is this *light* or *heavy*?" When a kid answered, "It is soft," I knew most of my words had likely gone over their heads. Lesson learned, my presentations to younger kids was revised.

More lessons learned, I left Kansas City, southwest bound. Lawrence, Kansas, was waiting.

Harnessing Science

For the first time in over two months, I strayed from my northward march. Backtracking through fields and suburban outcroppings, my indirect route veered southwest from Kansas City to the college town of Lawrence, Kansas (Go Jayhawks!). There I went in search of the monarch mecca known as Monarch Watch, a scientific and educational organization dedicated to understanding and protecting the monarch migration. Founded in 1992 by professor, entomologist, and monarch celebrity Dr. Chip Taylor, Monarch Watch is synonymous with monarch conservation. Knowing this, I had sent Chip an email while I was planning my trip, only half expecting a reply.

More than a reply, my email had led to a visit the summer before I left for Mexico. For several hours, Chip had advised me on my route and monarch science. Discussing both well-understood science and concepts he was only beginning to understand, he spoke as if I was the first person he had ever talked to about monarchs. There was no boredom in his voice, no patronizing explanations. A translator, messenger, conductor, he was the voice of the monarchs.

That visit lent credence to my guesswork and reassurance to my ambitions (though later he confessed to me that he had been a bit skeptical). I had been inspired by Chip's passion; he had spent a lifetime of work doing the daily chores of saving the monarchs.

Though the distance between Canada and me stretched wider as I ped-aled in the wrong direction toward Lawrence, I knew a Tour de Monarchs wouldn't be complete without a stop at Monarch Watch. As fortunate as a tailwind, my arrival (some 2400 miles from the overwintering grounds) had a touch of serendipity. I arrived just in time for Monarch Watch's annual spring open house and plant sale. This was like the Super Bowl for monarch fans: a day to buy plants, talk science, and rally together. A nod from the universe, my timing put me in the supportive company of many fighting as one.

At the open house, monarch lovers mingled. Adorned with a newly eclosed (just-emerged) monarch clinging to his short white beard, Chip greeted visitors, answered their questions, and, like a coach, encour-aged everyone. Angie Babbit, my host for my overnight visit to Lawrence, painted monarchs and caterpillars on kids' faces. She was in her element as both the communications coordinator and an artist. Sandy Schwinn and her friends from Tulsa came to buy native plants. Mary Nemecek (a con-servation superhero), Sam Stepp (a friend from high school), and other familiar faces from Kansas City came to celebrate the monarchs. Jacqui Knight came from New Zealand to speak of her efforts to conserve but-terflies and moths in kiwi country. In the luxurious garden, swallowtail caterpillars gorged on dill leaves. The yellow suns of white-laced daisies basked in their own light. Monarch pupa hung from potted plants and whispered secrets to every person who knelt to their level.

Along with the familiar faces were people I didn't know, but who were not strangers. I knew them, even without knowing their names. They were the team, inspired by a man, inspired by a butterfly; many paths led toward our goal.

The spring open house was the icing on a conservation cake. The rest of the year, Monarch Watch organizes two influential programs that combine science, conservation, and education: monarch tagging and way stations.

Each autumn, Monarch Watch enlists the help of volunteers across the continent to tag monarchs. Volunteers buy sheets of small round stickers, each with a unique combination of letters and numbers, a bit like a license plate. The tags are placed on the outer discal cell (the orange mitten-shaped block) of the monarch's hindwing. The monarch's location and date are recorded and submitted to Monarch Watch. If that butterfly is found along the migration, its path can be added to the growing database.

Even in Kansas, when I still had reservations about tagging, I saw reverence in such measures. Yes, tagging is scientific inquiry. It asks each butterfly, Will you make it to Mexico under the circumstances we have defined? Yet, at the same time, it feels like an act of faith. It tells each butterfly, I believe you will fly to Mexico. Tagged monarchs are both blessings and experiments released to open skies.

From 1992 to 2020, about 2 million monarchs were each given a unique identifying tag, and thousands were recaptured or recovered somewhere along their route. While many of these recaptures were found just miles from where they were originally tagged, the latest reports are that more than 19,000 tags have been recovered in Mexico. With each tag, the collection of data points grows bigger, and Chip sifts through the language of numbers looking for patterns.

Too much of the time, science ends here. Revelations backed by statistical probabilities stay hidden in publications and journals translatable only by those most fluent in their fields. The result? Implications smothered. Jargon unable to have impact. What makes Chip Taylor's story so powerful is that he did the opposite. He invited the public to become citizen scientists, and with the data they collected and the findings the data produced, he did something. In 2005, Monarch Watch launched a program to create monarch way stations. The goal is to encourage people to plant their yards and gardens with native plants and milkweed, allowing passing monarchs a place to eat, rest, and lay their eggs.

Whether you call them way stations, stopover sites, pit stops, or rest stops, travelers don't need an explanation of their importance. Like stepping-stones, they make the long journey possible, turning distant lands into homes away from home. As of May 2020, Monarch Watch had registered 28,210 way stations—28,210 plots of land to which the monarchs are invited, 28,210 ways to escape increasing evictions, 28,210 solutions to the problems identified by science.

While tagging benefits science and way stations combat habitat loss, it seems to me that Monarch Watch's greatest contributions, and Chip's by extension, are the connections formed by such direct action. Each of the 2 million tagged monarchs forms a connection to humanity. Every milkweed planted in the 28,210 way stations connects a gardener to the earth. With action comes solution. With action comes connections—connections to a team that is growing bigger every year, and to a migration that is growing smaller but that we will not give up on.

I looked for this connection in every monarch I saw. I wondered about each butterfly's life long after its buoyant body drifted out past the tide of sun-soaked clouds. Did this one visit a school, and inspire a young scientist? Did that one grow up in a yard, on the first milkweed someone had ever planted? Did another lay eggs along the road? Were those eggs mowed down, or did I pass them as caterpillars? What I did know was this: they had passed both rich and poor, rural and urban, Republicans and Democrats, and people of every color. There is only one sky and it is quilted by many migrants.

The monarchs made me wonder about the six degrees of separation theory. How many degrees would it take to connect Chip to every monarch? Watching him go from visitor to visitor at the open house, like a butterfly fluttering from flower to flower, I knew it had to be less than six. He bridged the gaps between science, conservation, and education. His work was also sounding the alarm bells of science so that real change could happen, instead of having distressing data simply fall into the void.

Seeing the effects of Chip's efforts was both inspirational and relieving. After many seasons of tracking amphibian declines as a field biologist, I had begun to see my role as clerical. I walked through streams that had once been so full of tadpoles and springy frogs that the waters seemed to boil, but now ran calm. I wondered how knowing such absence could do anyone any good (frogs included). I felt like a recorder of destruction, a bystander. Chip's work confirmed that scientists can discover problems, forge solutions, and make waves.

I hoped that, like Chip, I could dedicate my whole self to even the smallest of creatures.

Hope
in the Corn

Back on course after my short Monarch Watch detour, I found the route north from Kansas well established by both humans and non-humans. The Central Flyway that funnels most northbound monarchs is also an important passageway for migratory birds. Bald eagles and snow geese join buffleheads and goldeneyes in retracing ancient paths to sunny summers. The Missouri River carves its own trail by way of gravity, inviting mountains to visit the prairie, and parts of the prairie to head to the sea. Like Lewis and Clark (if they'd had bicycles), I followed, counter-current. Interstate Highways 29 and 35 are corridors, too, shooting northward like arrows.

I felt the importance of these passageways. You can't protect just one aspect of a traveler's journey; to protect the traveler, you must protect their every step, every wing beat. Migrating animals need safe habitat from here to there, in the summer, spring, winter, and fall. This large range of essential habitat is broken for most migrating creatures, making them vulnerable.

Conceivably, interstates could hold a piece of the solution. State departments of transportation and landowners surrounding the Interstate 35 corridor are partnering to restore native habitat on the lands they manage. A potential "monarch highway" could be grown alongside the paved highway. Though this habitat might not be ideal (scientists are studying the effects of traffic noise, winter salt applications, and toxic run-off on roadside habitat), it is certainly better than nothing. At an I-35 overpass, I stopped to stare down at the thoroughfare's wide, expansive ditch, mowed to the ground. It didn't have to be this way. Milkweed could be planted and mowed after the monarch migration. If only monarchs could call government officials.

The best thing about biking through corn country was its grid of farm-to-town roads. Navigation was easy and traffic was sparse. Aside from the occasional passenger car, I shared these back roads mostly with lumbering farm equipment. I braced myself for each passing hunk of metal, spikes and churning cogs cutting through the air mere feet from me, like beasts straining at an invisible leash. They engulfed the road, but they were slow and made sure to give me room as they passed. Slowly but surely, their hulking forms shrunk until they were consumed by the horizon.

Such machines were frightening, not because we shared the road, but because they were on their way to score huge swaths of land, land now unrecognizable to the monarchs. Humans had converted nature's breadbasket into a corn field. The monarchs' path was becoming barely a memory. On my bike, the miles bled together. Sometimes I felt that only my sorrow and corn stalks grew.

There were signs of life, opportunists at home in the rural skies. Meadowlarks used the miles of cable strung between utility poles to toss their whistles and warbles into the air. In cattail-choked ponds, confident red-winged blackbirds trilled and puffed. My passing was always met with a high-pitched *chek*, which drilled into my head. I wanted to tell the blackbirds that I was the least of their worries. Their homes were being erased by corn and lawns.

My ability to ignore the plight of the prairie began to erode, and the scope of destruction began to crush me. I looked for hope between the corn stalks.

At the Loess Bluffs National Wildlife Refuge off Interstate 29 in Missouri, I made eye contact with the pond turtles and blooming irises gazing across protected wet meadows. On the sixty miles of Wabash Trace bike path, I enjoyed the company of a startled bobcat. A forgotten highway ditch hosted a lonely milkweed. Pockets of wild calmed my anger.

I had a long line of people to thank for that remaining wild land, where civilization's clatter was hushed. Some of the people I knew, others were echoes. I walked quietly though their successes and let my footsteps give thanks. Heel touching earth: thanks for your foresight. Toes pushing away: thanks for fighting apathy with passion.

The best medicine against my anger was sharing my vision with students. I didn't hesitate to add miles if they led to a school visit. One such opportunity neared. Months before starting my trip, Kate Rezac, a third-grade teacher, had extended an invitation via email to visit St. Mary Margaret Elementary School in Omaha.

The extra miles to Omaha stretched a little longer, though, when the sky clapped with thunder a dozen miles from Kate's house. Lightning skeletons danced, and another storm, one of many in a long line that week, let loose. As uninvited as the rain, I ducked under an awning of the Driftwood Inn to wait out the weather.

Huddled in the cold, I willed the rain to relent. When that didn't work, I let my curiosity pull me toward the door, only to discover that the so-called inn was a bar. Not just any bar, the Driftwood Inn was an establishment with character, where beer was sold by the bucket and raffle tickets were currency. At four in the afternoon it was exactly what I would imagine a Midwest bar to look like at four in the afternoon. Eight or so customers flanked the actual bar, which stood like an island among inside jokes, handmade signs, and a few televisions. The sound of rain pounding on the pavement outside was faint. I ordered a Dr. Pepper, and was welcomed to the mix.

Slowly at first, the questions came. As I answered, the assembled faces went from half interested, to fully interested yet dubious, to shocked, before settling on a state of admiration. The bartender, who had diligently been making sure my Dr. Pepper was never more than a few inches from full, switched on the popcorn maker so I could have an equally unending snack. As the questions began to fade, my new friends turned their efforts to the rarely used oven, figuring out how it worked so they could cook me a frozen pizza. I assured them that I was happy with nothing more than a warm place to sit and their welcoming manner, but they assured me I needed pizza.

To pay for the pizza, a growing mountain of bills gathered on the bar. The bartender would accept none of it, pushing it all in my direction. I had been adopted by the folks of the Driftwood Inn, blessed by the magic of bicycle touring. Full, warm, cheerful, and remarkably richer, I left the Driftwood Inn, biked over the river (also the Iowa-Nebraska state line), and found myself in the maze of Kate's residential neighborhood.

My experiences at the Driftwood Inn and Kate Rezac's home felt very much the same: two families welcoming me into their lives, graciously providing me more than I deserved, and turning my trip into more than a bike ride.

I spent a few nights in the organized chaos of Kate's house. Juggling jam-packed schedules, Kate, her husband, and their three talented teenage daughters were busy. Only their wandering rabbit seemed to have any free time. Yet, as busy as they were, they somehow made time for me. On my first morning, long after everyone else had left for the day, I came into the kitchen and found a note from Annie, Kate's youngest. "Good morning Sara! We hope you slept well. Choose any food you'd like. I can't wait to see you at school." I read her note with gratitude; her warm words were gold.

At school, I felt that same positive energy and excitement from all the students. It was contagious, and as the kids filed into the gymnasium for my presentation, I watched them try to hold back their eagerness. Smiles spread as I waved, and surprised faces did double takes. In preparation for my visit, they had watched videos I'd been creating on the trip. Now it was clear that the lady they'd watched was not only real—she was in their cafeteria. So was her bike and her tent. Questions abounded.

"Why is your tent set up?" one first grader questioned.

"Because I am going to take a nap," I replied, watching him giggle at my joke. Ice broken.

Once everyone was organized in general rows, my presentation unfolded quickly. The kids knew more about monarchs than many of the adults with whom I spoke. They knew the whole monarch life cycle. They knew monarch caterpillars only eat milkweed.

In the blur of my rock star reception and the students' enthusiasm, I launched into an unscripted, ridiculous, and totally accidental blunder.

"I say MILK, you say WEED," I prompted. "MILK," I yelled before turning the microphone to the packed cafeteria.

"WEED," hundreds of kids in school uniforms shouted.

As soon as I heard it, I recognized the unintended alternate interpretation of what we were cheering for. I wanted to stop. Frantically running through options in my head, I decided an abrupt stop would only bring attention to the bizarre moment. I continued.

"Milk! . . . WEED! . . . Milk! . . . WEED! . . ."

No one looked horrified, and I moved on quickly.

Milkweed and the monarchs were not foreign to the students at St. Mary Margaret Elementary because they had been tending a butterfly garden and making real-world connections. I was treated to a tour of their garden by Kate's third grade class. They knew the plants well, and pointed out their discoveries as we walked the mulched paths. The whole school had recently united, class by class, to put down a new layer of mulch, and I felt the pride the children carried in their buoyant footsteps.

Their success had been hard won. The school had only reluctantly allowed Kate to start a garden. They had conceded a slope of grass that edged the parking lot, a plot that was difficult to mow and otherwise unnoticed. But with the help of Kate, her students, shovels, and some native plants, that patch had been transformed. It was now a classroom, a laboratory, and an escape from fluorescent lights and stale air. The garden had become an opportunity for classes to explore, for the community to come together, and for the monarchs to be teachers.

A television crew and reporters joined us to document the garden's success and my visit. I answered a few questions, but what I really wanted to do was explore. At the first chance, I let the students steal me away.

Even in the bluster of youthful energy allowed to roam free, I felt something akin to peace as we huddled around a milkweed plant to discover the garden's secrets. The milkweed, poked and prodded, began to ooze tiny bubbles of white latex sap. The tiny drops at first confused the kids. "Eggs!" they exclaimed, wild with excitement. "Look closer," I countered. It was not up to me to correct them. The garden was their teacher. "It's sap!" they squealed, delighted to have touched poison. As the group dispersed, shouting erupted. Real monarch eggs had been discovered by the young scientists.

Every day of my trip was an iteration of such learning. I could have studied thick research books on monarchs or read a hundred field guides on milkweed, but crouching in their world, trading carbon dioxide for monoxide, I learned real lessons that I would not forget after a test or a summer break. My senses wrote pages in my mind that could have filled textbooks. My curiosity turned the pages.

"A monarch!" a group of third graders exclaimed, their fingers and eyes tracking the sky. Like an echo, the whole class joined to watch the monarch float over our heads. Their pure delight and his delicate orange wings filled my heart with hope. Human beings were not made to fight nature. Human beings are made of nature. When we are released to waltz among gardens unburdened by the appetite of power, we can see our home in nature—and see that we share our home with monarch butterflies.

That monarch was a messenger. It christened the air with wing beats and ordained each student as a scientist, conservationist, and member of our shared planet. That monarch connected the students to Mexico, Canada, and New York City, not just an unnamed patch of schoolyard in Nebraska. He had navigated an unwelcoming ecosystem of grass, pavement, cement, brick, tile, plastic, shingle, and wood, and found refuge in the class's small garden. That monarch was proof that when we give nature room, it will thank us with color and so much more.

I stood, as stunned as the students. Until a monarch finds you, until your paths cross, the astonishment you will feel can only be a word on a page. Until every kid feels that connection, we are failing them.

I began to envision a garden at every school, providing every student with a living laboratory and real-life connection with nature. The monarch could become a symbol of our shared responsibility, shared planet, and the power of our combined efforts. With a native butterfly garden at each school, it wouldn't matter where the kids lived, how their state representatives voted, what they looked like, how they talked, or how much money their families made. The monarchs could connect us all, reminding us that we are all creatures that deserve to be seen as amazing.

Kate's determination and courage to start a garden that had first been met with rejection was inspiring. I needed her garden as much as the monarchs did. I needed to see that there were people doing their part, helping steer the future toward something better. Her dedication to making sure the garden remained a success was welcome fortification for me.

Crossing the Missouri River again, I moved north. I crossed corn fields and a state line into Iowa. At Desoto National Wildlife Refuge, layers of native grass hid speckled fawns, and marshlands were flooded with pipping kingbirds. A view of a raccoon, noble in its wildness, buoyed me. I crossed more corn to arrive in Mondamin, Iowa (population: 379; named after a Native American goddess of . . . you guessed it, corn). Near the playground of the local school, middle schoolers walked me through a corner dedicated to native prairie. Next up was Sioux City, Iowa. On a meadow-topped hill overlooking town, a ringed-neck snake said hello. Yet, between the hopeful moments were miles of corn. I could not isolate myself from the truth.

"We used to see loads of monarchs," a man at a grocery store mused. "Now I hardly see any." I'd heard this almost daily. It was a story told without malice, but it stung all the same. Some of the stories I was hearing, I enjoyed. Such as how, in World War II, the government paid children to collect the fluffy seed heads of milkweed (called floss, or coma) in onion bags. The floss went into making life jackets for soldiers. It took two onions sacks to fill a life jacket, and 1.2 million jackets were needed. Milkweed gave us over a million life jackets, and our thanks was to spray it with weed killer.

Mostly, the stories just hurt. The older generation had lived with milkweed when it was still abundant (and unappreciated) enough to be called a weed. The people sharing their stories would talk about how, as kids, their jobs had been to pluck milkweed from the fields, but their stories mostly ended there. They didn't usually continue with how, as they became adults, their family farms were bought out and mega-industrial agricultural conglomerates began to rule the land. They didn't share the fact that, by the 2000s, genetically modified crops and broad-spectrum herbicides had been unleashed. Their childhood farms were replaced by fields of modified corn, which became the only plant capable of withstanding the poison doused on every acre. Even milkweed, once prolific, largely succumbed. The last holdouts of the prairie didn't stand a chance. Only gene-altered corn survived, creating naked and poisoned monoscapes.

That older generation had witnessed something spectacular, then they had bequeathed me a shadow. Most never asked themselves why or said sorry. What are you doing to fix it? I wanted to ask, but my diplomacy usually won out. "Current farming practices don't leave room for native plants like they once did," I explained. "Monarchs *require* milkweed. Without milkweed, there can be no monarchs. We have to change how we farm." After such conversations, I hoped, as I biked away, that my words would leave them unsettled, and they would look around and begin to tell a different story.

There were once billions of monarchs, now there are millions. There had been millions of Eskimo curlews, billions of passenger pigeons, and trillions of Rocky Mountain locusts—now there are zero.

These animals are trying to warn us. The milkweed and the monarchs are canaries in the coal mine. Humans, numbering nearly 8 billion, are also threatened. The broken planet that is killing insects and birds and frogs is killing us, too. We are under attack. Watch the news. Go outside. A report by an Australian think tank ran the numbers: if nothing changes, human civilization as we know it today will likely collapse by 2050. I'll be 65. I hope I die before the monarch migration does.

I am aware how dramatic that sounds, but my heart can't stand the pain of telling a generation not yet born that when I was young, there were still monarch migrations.

I am also aware that we must eat. That we must balance farmland and protected lands. It is a matter of priorities. It is about remembering how to share. There was a time when plowed farmland actually invited common milkweed (*Asclepias syriaca*).

Prior to European settlement, common milkweed had been a species with a limited range. As a disturbance species, it was likely confined to excavated soil, mostly found around animal burrows. Plowing opened up millions of acres of opportunity. Common milkweed thrived between corn rows and on field edges, and was actually preferred by egg-laying female monarchs over adjacent nonagricultural habitat. This preference may have been because such milkweed was easier to find, occurred in isolated patches which limited predators, had more edible leaves (due to lower sun exposure), or had higher nitrogen levels. Whatever the reason, it is good to know that sharing is possible. Today we don't often share, even accidently.

Herbicide use began in the 1940s and was historically applied before the corn emerged; otherwise the corn also would have been killed. Glyphosate (Roundup, as marketed by Monsanto) was introduced to control "weeds," but again, it could not be applied once crops had emerged. Spraying before emergence allowed monarchs to safely utilize agricultural habitat.

In 1996, genetically modified (GM) soybeans were introduced; GM corn followed in 1998. These seeds were tolerant to glyphosate applications, and being "Roundup ready" meant that farmers could spray after the crops had emerged. By 2012, 73 percent of corn and 93 percent of soy were Roundup ready. Any plant that was not tolerant quickly starved, as their mechanisms of photosynthesis broke down.

Milkweed didn't stand a chance.

In 2019, the United States Department of Agriculture reported that more than 91.7 million acres (about 69 million football fields) of corn were planted in the nation, and 89 percent sprouted from herbicide-tolerant, genetically modified seeds. Our water, air, milkweed plants, and monarchs continue to be collateral damage.

As I watched the miles of corn come and go, the most infuriating part of the puzzle was that most of that corn wasn't even being grown for human consumption. Depending on markets, about a third of it goes to animal feed, another third to ethanol production. Ethanol is a fuel source that uses more energy to make than it can produce, yet the government supports it with subsidies. We call government support to farmers "subsidies." Support for poor people is instead referred to as "welfare." Regulations are basically subsidies for wildlife and wildlands. I yelled at the clouds, because they listened. I wished they could tell me what they remembered of that place once called a prairie.

On a cool morning a few days north of Sioux City, my frustration overflowed as I spoke with a reporter over the phone. "Are you always this angry?" he asked. His question, an innocuous dismissal of my concerns, was a punch to the gut.

Was I always this angry?!

Before starting my bicycle tour, some 3000 miles earlier, my knowledge of the monarch's plight had been mostly academic. I knew that in the last twenty years the monarchs had seen a dramatic decline in their population, a decline linked to habitat loss across their range. Such theoretical truths had fueled irritation and frustration in me. Now, mile after mile, the endlessness of corn and sprawl had become consuming reminders of what was no longer.

I was using my body and time and voice to speak for the monarchs, absorbing the truth of their loss. The reaction to my outrage was not to question the system but instead to judge my inability (or unwillingness) to stay calm. Stay "rational." Stay polite and ignore the rumble of devastation. Margaret Murie, a passionate conservationist, understood the importance of not letting cultural norms dilute concern. When she testified to the US Congress on behalf of wilderness in 1977, she told the politicians, "I am testifying as an emotional woman, and I would like to ask you, gentlemen, what's wrong with emotion?"

Murie insisted that "Alaska must be allowed to be Alaska," and her passion helped protect a piece of wilderness I may never see. Knowing it exists, however, sustains my spirit—though renewed attacks on the Arctic National Wildlife Refuge continue today. The next generation will want wilderness, too. They will want places that are less broken to connect them to the soul of their planet, where the past and future dance more peacefully together.

The monarchs must be allowed to be monarchs. I say this as an emotional woman, and I ask you, everyone, how can we afford to be anything less than emotional?

The reporter's question festered.

I *was* angry. I was angry because I was being told to accept a broken planet, because fixing it was inconvenient. Because fixing meant change. Because change was hard. I was angry because the powerful labeled opposition as overreaction.

I *was* angry that I was nearly 3000 miles into my journey, biking through the historic heart of the monarch migration, and I wasn't seeing monarchs. I wasn't seeing milkweed. I was angry that we had sacrificed a rainbow of life by planting poison.

I *was* angry that I was seen as melodramatic. Angry that I had to tiptoe around my pain. I was angry that the unwritten rule was that composed apathy and status quo were dignified, while speaking the truth and calling people out for their complacency was unbecoming.

Was anger inappropriate? To me, only utter rage was more appropriate.

"Not always," I replied to the reporter. Even as my anger boiled, I knew that erupting would not work. I still had to play by his rules.

I told him about the moments when my anger felt far away. Times when the symphony of rustling plants, bellowing birds, and the nearly unbroken wind reminded me there was still much to save. I told him about being surrounded by children at their school pollinator gardens and the strength I felt as they shouted with joy at a passing butterfly.

I told him about the people I had met who were planting, teaching, promoting, supporting, growing, leading, encouraging, and nurturing—all to help ensure a future that held visits from monarch butterflies. I told him about the older faces that knew what had been lost and the younger faces that knew what had been stolen.

Was I not allowed to be angry? Was I naïve to have hope?

I didn't know the answers. I said goodbye, hung up the phone, and carried on north. Minnesota was just a few days away.

Spring to Summer

At the border of Iowa and Minnesota, I ate an apple in the windbreak of the Land-of-10,000-Lake's welcome sign. A monarch, the first I had seen in several days, settled into the neighborhood of dandelions at my feet. He inspected several of the yellow flowers with his proboscis and feet, a dance of touch and taste. I joined the monarch, settling into the flower's beauty. Why do we shun those little yellow suns?

Full of dandelion nectar, the monarch drifted into the stream of wind, hesitating for a second before powering toward another tempting flower. I too powered into the wind, into Minnesota. Spring had turned to summer, and a galaxy of yellow and white flowers grew in the space along the road. Heavy clouds blocked the sun but did not dull the flowers' bright petals or my relief at finding nature.

Advancing against a determined wind, I was also relieved to find a town park not far from the border. I often took breaks in city parks, but rarely did I find such luxury. Boasting a shelter to break the wind, a spigot to fill my water bottles, and working electrical outlets, the park was a bicycle touring miracle. I quickly went to work transforming it into my kitchen, office, and lounge.

I was halfway through my vegetable wrap and sorting out my schedule when a group of kids, freed from school and shoes by the summer holiday, rolled up on bikes, cursing loudly to prove their coolness. I ignored the cursing but not their questions, and they quickly uncovered my story and inspected my bike. All attempts to be tough yielded. I was transformed from a stranger to a custodian of possibility. Their questions, like all questions from young people, revealed their minds processing the potential. You carry a tent so you can sleep anywhere? You carry a stove so you can cook anywhere? Anywhere? You can bike from Mexico to Canada?

To ready them for their own adventures, I helped fix a flat tire and put oil on their bike chains. They wished me a safe journey, an octave higher but with the same enunciation their parents would have used. I waved goodbye and headed back to the wind.

The grocery store at Albert Lea, Minnesota, offered my next windbreak. The cement-block walls—lined with an array of soda machines, newspaper stands, and abandoned grocery carts—braced the wind as I stuffed, squeezed, and tied down the overflow of my purchases. Shopping while hungry had led to a cart of pickles, nori sheets, apples, gummy bears, sunflower seeds, lettuce, salad dressing, bread, and pretzels. It was going to be a strange feast, but as soon as I found a place to camp, it would be a feast all the same.

I pedaled two miles out of town and found the church I had reconned on Google Maps. Rural churches usually made for great Midwest camping spots. They were tucked away, with quiet corners, lazy hours, welcoming philosophies, and the confusion of no primary tenant (Mary thinks Jim probably gave permission, Jim thinks Mary probably gave permission). This time, however, the church, surrounded by houses, was not a feasible option. Before I could make a Plan B, the universe raised its cosmic eyebrow. My phone rang with a call from Tom Ehrhardt.

I had met Tom at the grocery store, and originally declined his invitation to stay at his house. I knew I was breaking my say-yes rule, but right then, I didn't think I had the energy to navigate a stranger's world. The miles of corn had taken their toll on my spirits, and I felt buried by inertia. Perhaps when Tom called the number on my business card to check in, he knew this. Given a second chance, I said yes.

Tom's wife, son, and two dogs greeted me as I pulled into the driveway of their quaint house. It was encompassed by feral land, richly diverse roadside ditches, and an area he'd recently converted to native prairie. I knew I had made the right decision. After being bogged down by the reality of our corn-covered prairie, meeting Tom was a gift. He owned Albert Lea Seed House, and explained that markets and math were helping farmers return prairie to the planet. Instead of losing money when yield prices didn't cover costs (corn was losing $100 an acre, Tom told me), farmers were finding alternatives. For example, farmers enrolled in the USDA's Conservation Reserve Program receive a yearly rental payment to remove environmentally sensitive land from agriculture production and plant species that will improve environmental health and quality. Each contract is between ten and fifteen years long, giving the land time to improve water quality, prevent erosion, and provide wildlife habitat. "Every day," he assured me, "farmers call to ask me what they can plant besides corn." He fed me the hope I needed as much as food.

I left Tom's feeling much better than I had when I'd arrived, and I carried that renewed faith and a large quantity of milkweed seeds, gifts from Tom, northeastward into Rochester, Minnesota, on a milkweed-lined bike path. There I stayed with a friend, Andrew Schmid, who had extended an invitation when we had first met on the face of a mighty Sierran mountain the previous summer. In his basement, now far from that alpine setting, we watched the nightly news and the interview I had done that day.

The past unrolled into the present with a slight sense of the surreal. Surreal but not surprising, though. Traveling teaches us that where we go is often decided by the collection of people we have met and places we have been. I went to Rochester because I'd met Andrew, because I'd gone backpacking in the Sierra, because I'd fallen in love with mountains, because I'd worked in Glacier National Park. Because, because, because.

Because I stayed with Kate in Omaha, I was invited to stay with her sister and brother-in-law in the Twin Cities. Because I was headed from Rochester to the Twin Cities, I chose a route north, on a residential road. Of all the possibilities, that route and timing led me to connect with two orange and black butterflies. I stopped to watch them twirl in the air like synchronized acrobats. Though I had first stopped, thinking they were monarchs, at second glance I identified them to be copycats. It was my first sighting of the famous viceroy butterflies (*Limenitis archippus*). Except for the viceroys' arc of black that divided each orange hind wing and their more organized, white-dotted wing edging, I understood why animals, including humans, might confuse them with monarchs.

For some predators, the mere fact that a potential target is orange is enough to keep the predator at bay—what's known technically as aposematic coloration. Predators are wary of orange, a caution owed to experience and evolution. When the warning colors of monarchs are not heeded, all but a handful of vertebrate predators will learn at first bite that the orange butterflies are toxic. This defense is a result of monarch caterpillars eating milkweed and sequestering the plant's cardenolide toxins in their bodies. The cardenolides are carried into adulthood, making adult monarchs a bitter emetic. A clueless predator will often release a monarch after one bite, the only damage a chunk of missing wing. Once the lesson is learned, anything resembling the cause of such discomfort is usually safe, just as humans often avoid a food associated with a time when they were sick.

The viceroy is a mimic. With a wing color and design similar to the monarch's, it can piggyback on the monarch's poisonous protection. Predators can't tell them apart, and so both are avoided. There are several classifications of mimicry. The viceroy was for a long time considered a Batesian mimic, in which the monarch was the unpalatable model and the viceroy was palatable. In this system, if a predator first eats a viceroy and suffers no ill consequences, then the monarchs are actually more vulnerable because predators will think they are safe to eat as well.

In the early 1990s, though, the Batesian classification was called into question. It seems that viceroy butterflies are also unpalatable. Viceroy caterpillars eat Carolina willow (*Salix caroliniana*) and, like monarchs, sequester the plant's defensive compounds. As adults, viceroy butterflies secrete these defensive compounds when disturbed, deterring predation.

Mullerian co-mimicry, in which the viceroy and the monarch are both unpalatable models, is now deemed a more appropriate classification.

A couple walking through the viceroys' neighborhood watched me as I ran from one plant to the next, my camera trained on details invisible to them. Until we learn to see, we may never be able to distinguish between orange wings, green leaves, and blue skies. We may continue to be lost even in the places we call home.

Under the spotlight of attention, the viceroys steered into the sky. I left on my bike, following forty miles of flower-framed bike path. I witnessed several pairs of monarchs breeding. The cottonwoods celebrated summer with confetti of seedy fluff. As the day wore on, I began to catch glimpses of Minneapolis and Saint Paul, which hovered like distant clouds. The cities were still too far to reach before dark, so I began to look for a place to stay the night.

On a scouting mission, I followed the long driveway of a Baptist church and found exactly what I sought. Beyond the grass, behind the brick wall, alongside the air conditioning unit, was a spot nearly hidden from the view of passing cars. To the west, shrubs stood guard. To the east lay a naked, unplowed field. I set my bike down and stretched with satisfaction. Another nearly forgotten space had just become my home.

As the day awoke, so did I—my alarm clock, the low rumble of a tractor sauntering through the field that butted up against the church. After adjusting to the light of a new day, I waved to the man in the driver's seat. I figured a friendly wave would lower suspicion. Assuming he was too busy to do anything about the random lady in the tent behind the church, I went back to sleep.

The next time I opened my eyes, a few hours later, it was to a lawnmower cutting the grass literally around my tent. The guy atop the riding lawn mower treated me like the playground equipment, and I treated him like a more adamant alarm clock. I packed up without pause, waved yet again, and biked back to the highway. I wondered as I biked how that interaction would have gone if I had been a black man. Would the police have been called? Would I have been able to bike away?

This question of privilege surfaced again some fifty miles north of the Twin Cities, as cops shook my tent and yelled commands to wake up. I had chosen the camping spot in a park with access to a forested hiking trail, even though there was a No Camping sign. Normally the sign would have deterred me, but I had been caught on a highway constricted by construction. Continuing in darkness could have been deadly. Instead of slinking guiltily into the woods and potentially scaring an early morning visitor, I camped in plain sight, my bike propped up to explain my traveler's dilemma.

This did not stop the cops. "Get up!" they shouted as they pounded on my tent. "Police!" Jolted awake, I was in fight mode before my brain could catch up. I answered with a startled string of cuss words. Then, completely awake, I unzipped my tent and handed over my driver's license. They hovered over me and grilled me with questions. The degree of their anger did not seem to match the degree of my crime. Luckily it was 7 a.m., so I was happy to get the heck out of there when they told me go. I understood it was their job to enforce the rules, even arbitrary ones, but did it need to be done with such venom and lack of curiosity? I had pitched my tent out of the way, I had picked up the nearby litter, and all I was doing was sleeping.

My blood boiled as I packed up, and I wondered how it would feel to live in a world where it was deadly to demand dignity. Only months earlier, not far from where I'd camped, Philando Castile had been shot in his car by a police officer. His girlfriend, Diamond Reynolds, had sat in the driver's seat with a calmness that I have never been forced to muster. Her baby was in the back seat, Philando was dying at her side, and she called the murderer of her boyfriend "sir," while he pointed a gun at her.

I had broken a law. Philando Castile had a broken taillight. I had been allowed the space to be indignant. Diamond Reynolds couldn't afford that. What if the police had found her in my tent? What if her boyfriend had awakened surprised and cussed in alarm? It was difficult to accept that my skin color was in part what made my trip feasible. I questioned how I could prioritize monarchs when injustice was swinging its deadly sword at my human brothers and sisters of color.

I began to see the monarch as a symbol of compassion. No one blames the monarchs for their population plummet. No one says it is the monarchs' fault when caterpillars die from herbicide and pesticide use. No one thinks that monarchs are lazy because they can't find milkweed anymore, or that they deserve getting hit by a car because their migration passes over Interstate 35. Instead, we recognize the monarchs' struggle. We rally, fight, cry, get angry, and do something.

We are capable of seeing that the world is an equation, tweaked for some so that the result is best for those making the rules and moving the pieces.

The monarchs were helping me to see many worlds.

A Summer Break to Bike

Until I reached Lake Superior's shores, my trip had been a summons north. Yet like the monarchs, I couldn't go north forever. I veered and let the Upper Peninsula of Michigan funnel me east. The change in direction was also marked by a change in my daily routine. With schools closed in the US for summer break, my tour became simpler. Instead of the logistics that came with presentations to students, all I had to do was bike—a lot.

My next presentation was scheduled in ten days, 700 miles away in Sudbury, Ontario, Canada. I would need to cycle through every wind, every rain, and every obstacle that came my way as I rushed across the arm of Michigan holding Lake Superior to the north and Lake Michigan to the south. The only thing I would let slow me down were the creatures living along my route. No matter how rushed I was, I would always make time to pause in their company.

Spotting a snapping turtle plopped on the side of a narrow OHV (Off-Highway Vehicle) path, I stopped, dropped my bike, and, keeping my distance, knelt until we were at the same level. She stared at me wearily, her small marble eyes filled with golden flames. Her beak was striped black and gold, matching her gaze. I knew she could leave me fingerless if I gave her the chance. Her smooth shell barely held her body and was covered in scratches that told the stories of her life. I could not leave her on the trail, vulnerable to the threat of careening off-road vehicles, but I couldn't exactly pick her up. I tried a friendly stick prodding, but she whipped her neck around and struck at the stick before hissing at me. Her fear translated as a fight.

"Okay, beautiful," I told her, a safe distance away. "I know you are scared, but I am not leaving you on this road." I grabbed her tail and dragged her backward, her muscles contracting and her power radiating to my hand as her fright turned to fury. Both of our hearts were pounding as we backed up into the brush. When I let go and leapt away, her marbled eyes burned into mine as her body sat as still as a heavily breathing rock. "Sorry," I said. "I owe you more."

That snapper was a mere preview of what was to come. Two days later, on another trail for all-terrain vehicles, I watched a wood turtle amble toward me, a muddy shell betraying its earthen home. A mile later, another wood turtle caught my eye. A ray of golden yellow stretched from ancient eyes down a leathery outstretched neck, to the cavern made by the shell. Four stubby legs, decorated with long, graceful claws, paused.

It wasn't until I passed a clique of female turtles, all in different stages of egg laying, that I put the pieces together. A recent summer rain had triggered the female turtles to leave the safety of their waters and lay their eggs on land. Mile after mile, I saw venturing mother turtles—snappers, woodies, and painteds—digging nests in the soft edges of the trail. A few were filling their excavated nests with eggs; others were testing potential spots for suitability. I scolded each one in danger of being run over, noting the shine of their shells, the details in their patterns, and the speckles in their eyes. I cajoled moms not yet committed to their nests to move farther off the road, and I moved brush and rocks strategically, in the hope of herding vehicles away from vulnerable eggs. A nursery in a trail's shoulder? Humans keep taking, and wildlife keeps trying to make do.

Turtles weren't my only distractions Along forest edges crept lanky sandhill cranes. A coyote scampered. Black and yellow knots of swallowtail butterflies combed the air and pooled on the gravel shoulders. East of Duluth, Minnesota, I approached a tree that was singing with the confident caws of conversing ravens. Yet as I got closer, I was startled to see white forms among the branches. My eyes struggled to translate. Ravens are black. In front of me were several black ravens (as I had always seen), and yet there were also two cloaked in white feathers. I snapped a photo to prove I wasn't dreaming. Time was pressing, but how could I not stop for mystic birds, egg-laying turtles, and a swim in the corrugated reflections of lakes as big as oceans? Well, I conceded, I wasn't *that* rushed.

Passing a field of lupine, I again had to stop. A bee hovered beside a purple bloom. On a milkweed, a monarch caterpillar ate its way to adulthood. On a neighboring plant, an armor-shelled insect was on the prowl. Though my insect identification was precarious, I called it a stink bug. With a piercing proboscis, a stink bug sucks the liquid out of its prey.

I gulped, but did nothing. This was both a trail and a place of wildness; the monarchs were in the thick of it. Despite being poisonous, monarchs are vulnerable to many potential predators. Spiders, wasps, ants, true bugs (the Hemiptera), and flies all prey on monarch eggs, larvae, and pupa. Using data from citizen scientists who tracked survivorship from egg to caterpillar, it seems as few as 20 percent of eggs become larvae, 10 percent of larvae reach second instar stage, and 2 percent survive to become third instar caterpillars. Though the exact numbers vary from study to study, it is generally agreed that less than 10 percent of eggs reach adulthood (likely much less).

I turned milkweed leaves in careful inspection, as if reading a book. A pack of aphids crowded the tips of budding flowers. While correlation and causation are often confused, there are interesting links being made between monarch predation and hosts. There is evidence that the abundance of critters on milkweeds influences the survivorship of monarchs. This is likely because prey (such as aphids) attracts predators, and some of the predators (like spiders) will be generalists. The generalists will eat the aphids and any monarchs they happen to stumble upon. The same idea plays out with flowering milkweed. The flowers attract pollinators, which attract potential predators, thus increasing the occasions for an opportunistic predator to find a monarch. Perhaps even the monarchs themselves are ringing the dinner bell. Milkweed damaged by herbivory may attract predators that key into either the defensive chemicals released by the injured milkweed, or by visual clues, such as white sap.

Predator-prey relationships are complicated dances. Each step is an important part of the whole: each creature, each meal, each escape, each doomed egg. If monarchs didn't have natural predators, a female could lay her 400 eggs and by the end of the summer, have 64,000,000 great-grandchildren. The monarchs don't need all their eggs to survive. What they need is enough milkweed, so they can lay enough eggs that at least a few of the eggs will beat the odds.

I was halfway across the Upper Peninsula of Michigan, back on a paved highway, when the storm clouds that had been casing the horizon for days enclosed me. Electrically charged thunderheads bullied the greening skies. The first drops fell delicately.

It was 6:30 p.m. and I briefly considered calling it a day, but my schedule was too tight to slack off. Besides, I liked biking in the rain. As the delicate drops grew into a frenzy, I bundled into my raingear, replaced my shoes with sandals, tightened the buckles of my panniers, and turned on my blinky bike lights. There was to be no escape, and I was ready. Sheets of rain began to fall from the doughy clouds above. "This," I shouted, soaked and happy, to the universe, "is what it feels like to bike in the rain." My end-of-the-day tiredness was lifted by the onslaught of water, and I biked with purpose, chased by the taunting storm. Soon the rain was replaced by the wet-whip sting of hail. My hands flinched as the hailstones nicked my skin. I tucked in one arm, using my body as a shield. Alternating arms, I laughed to the clouds. *Who volunteers for this?*

"I do!" My hoots were drowned out by the rain.

I stopped laughing when the road's shoulder disappeared and traffic picked up. Each passing car churned up a thick, watery wall, and I was uncertain if trailing cars could see me. It seemed like an unnecessary risk to carry on, so I biked off the road, across the ditch, and into the company of equally wet trees. I considered waiting the storm out, but darkness was descending and it was a guessing game as to whether the storm would last minutes or hours. It didn't seem to be letting up, so I plotted a strategy to get my tent up as fast and as dry as possible. Instead, my cold fingers betrayed me, and I set up my tent in the slowest, wettest way possible. The outer rainfly of my tent became a funnel, directing all the rain through the mesh and into my limp tent, as I bumbled in the storm.

By the time my tent was up and I was inside, I didn't know whether to laugh or cry. I told myself that tears would only add to the lake that was my floor, so grumbling, I set out to make the tent livable. I used my cooking pot to bail water out, strategically placed my pack towel to make an island of dry, and traded my wet clothes for dry pajamas. I wrung out my shirt and pants and used them as towels to wipe the last puddles from the floor, before throwing them into the tent's vestibule. I'd deal with them in the morning. Next, I unrolled my sleeping pad, pleased with my now warm, dry home. I dug through my waterproof panniers (sitting with my wet riding clothes in the vestibule), and rustled up a meal: a cereal-and-honey-butter burrito, followed by a sandwich made with bread, a mountain of lettuce, sunflower seeds, and some salad dressing (salad sandwich—no bowl required!). A handful of gummy bears were dessert.

The sky apologized the next day with warm rays that dried my gear and blessed a swim in Lake Michigan. I dipped under freshwater waves without tides. Unlike my frantic splash in Lake Superior a few days earlier, the water was warm enough for a prolonged soak, yet cool enough that I felt refreshed. The Great Lakes were like watery states—neighbors, but each its own. I made it a goal to swim in all five.

Like any good tourist, I also made it a goal to try the Upper Peninsula's signature pasty, a potpie you eat with your hands, and a throwback to both immigrant and mining culture. Today, pasties are little more than roadside distractions—but I like roadside distractions. I stopped twice for pasties and twenty-four times for twenty-four monarchs: a daily record for my trip. I also splurged on a ferry ride to Mackinac Island, an island without cars that lures tourists with fudge and roads filled only by the traffic of cyclists, horses, and pedestrians. Biking around the island, I saw three monarchs, flitting from flower to flower against the backdrop of Lake Huron's calm waters.

A tailwind pushed me north to the Canadian border, delineated at my location by the living, churning Saint Marys River. There I climbed the spanning, narrow bridge under construction. A line of cars grew long behind me. A crosswind whipped down the river valley. I gripped my handlebars and concentrated on staying my course. At the crest of the bridge, gravity took over. I picked up speed as I left the United States and was carried into Canada.

The North Lands

I scrambled my way through a labyrinth of one-way streets, exchanged US dollars for loonies and toonies, and had a photo shoot with a Canadian flag. Near Val Biron's house, where I was expected, I stopped to inspect a clump of common milkweed. Along a retired railroad track it had proudly reclaimed its space. On one fuzzy leaf perched a monarch egg. That egg was likely the great-grandchild of a monarch I had followed north from Mexico.

The multigenerational monarch migration can be broken up, generally speaking, into four smaller migrations. The first and fourth, when the monarchs fly purposely north in the spring and south in the fall, are the most obvious. The second migration is the continued movement north by the progeny of the monarchs that overwintered in Mexico. Born in the southern United States, this generation develops into adults that continue the migration north. It is not clear, however, if they are purposely orienting their flight north like the monarchs migrating from Mexico. Rather than using the sun as a compass, they may simply be following the northward emergence of milkweed or escaping warming southern temperatures (and laying eggs as they go).

Regardless, by early June the monarchs have recolonized their northern range, and directional flight ceases for at least several weeks. During most of June and July, adult monarchs stay localized. By late July, the third (also known as midsummer) migration begins, but it is experienced by only some of the monarchs. Typically a portion of this generation continues to live locally, and the rest carry out the third migration of the season.

This third migration is in essence a mini migration south—a precursor to the fourth migration that will reach Mexico. My best guess was that the egg, the first I saw in Canada, was destined to be either a local meanderer or a part of the third migration. It was impossible to know which. "Good luck," I said to the soon-to-be butterfly.

Though I was at the northern edge of the range, I was not fated to loiter. I had traversed the monarchs' Central Flyway headed north, but that still left the Eastern Flyway. Each year, to varying degrees, monarchs are pushed east by prevailing winds. I liked the thought of getting pushed alongside butterflies to the last state in the contiguous United States I had yet to pedal—Rhode Island—while also shining a spotlight on more of the monarchs' range. Stretching my route east to the Atlantic Ocean meant that I would pause only briefly, 200 miles to the northeast, for my next set of presentations in Sudbury, Ontario. I had three days to get there.

The eastern range of the monarch is a relatively new phenomenon. Just as the common milkweed's distribution changed with the arrival of European settlers, so did the distribution of milkweed species in general, and as a consequence, the distribution of monarchs. Historically, the monarchs were concentrated in the prairies of central North America, where milkweed was abundant. But since the 1800s, that milkweed-filled prairie had become farmland and the deciduous forests from the Great Lakes to the East Coast had been demolished. Milkweed, removed from its native home, found new holds in the once-forested lands. The monarchs had followed.

En route to Sudbury (now part of the monarch's range), I began the day on glorious backroads twisting through a patchwork of farms and forest. Mennonite farmers smiling under wide-brimmed hats and bonnets passed me in horse-drawn buggies. Country homes, barns, farming equipment, and time sat in piles along the roads. Milkweed framed the views and I took deep breaths with all my senses. I was unaware that hidden in that quaint landscape lurked the scariest wildlife encounter of my trip.

The views, the quiet, the serenity lulled me into a content rhythm. Up the hills, oblivious, I watched horizons unfold. First the crowns of trees, then the rusty weathervanes of antique farmhouses, and finally the expanded views of swirling countryside. Down the hills, I let my legs rest while the wind drenched my skin and the pavement became a blur. Up, down. Up, down. Up, down. *Whoa!!*

In the road teetered a very-much-alive skunk. I'd seen so many dead ones that the joy of finding one alive clouded my senses. Only when the skunk paused, startled, to question the air with her nose, did it dawn on me that she might not be as excited as I was about our meeting. I released my brakes, letting momentum and gravity glide me past the potential danger. A good distance away I came to a stop. I turned to watch and come up with a plan.

The skunk made its way along the edge of the road, looking for a way to cross the waterlogged ditch. I doubled back, passing her with a nervous awe, aware that I was close enough to see her nostrils flare with attention. Once out of the way, I stood as still as I could, and the skunk, still looking for a way across, unknowingly closed the distance.

With only the width of the road separating us, the skunk paused to sniff the gap. Her curious nose and wary eyes read the wind. The white stripes on her black fur matched the pavement markings, except in the chaos of her tail. My heart pounded; the worst-case scenario flashed in my mind.

Getting sprayed, with no shower in sight, and no change of clothes, I would smell like a skunk for days, months, maybe the rest of my trip. Someone had once told me that bathing in tomato juice was the antidote to skunk spray, but I only had ketchup. *Would ketchup work?* I imagined rolling up to my next school presentation on my junky bike covered in ketchup and smelling like a skunk.

"Worth the risk," I muttered as the skunk's beady eyes tried to figure out what I was. It is often the animals we think we know—the skunks, raccoons, possums—that are the most interesting. In the context of their wildness, their beauty can eclipse the most exotic creatures found in travel magazines.

I left the skunk when she finally found a route into the secrecy of thick grass. She didn't spray me. Looking back, I am almost disappointed that I escaped. Covering myself in ketchup would have made for an epic story.

I carried on and soon thickets of milkweed began to distract me. The common milkweed lining the road both confirmed my route and gave me hope. As long as there was milkweed, I was on a decent route. As long as there was milkweed, the monarchs stood a chance. I noted the confidence of the stretching plants, and took comfort in their familiar wave until reality kicked me in the face.

The roadside ditches crowded by milkweed ended abruptly, and I was confronted by a long stretch of road flanked by a short, green shoulder manicured by a recent mow. The tops of all the wildflowers lay strewn about. Still green, each plant was in denial, still living without a future. I stopped to inspect the beheaded milkweed, and saw what I feared. On each plant I picked up, sat an egg or tiny caterpillar. Dozens upon dozens of the next generation, unaware that their home had just been hacked.

I left my bike to collect the homeless eggs and caterpillars and relocate them to spared milkweed farther from the road. Sixty eggs later, I knew it was unsustainable to walk across Canada transplanting eggs. My limitations bogged me down. My frustration boiled. I wanted the cathartic release of accusation, and a mile down the road, I spotted an unsuspecting target for my despair. A man sat on his porch swing, happily unaware that the recent mowing had left some of his neighbors to die.

The tiny eggs in my hand, accepting their fate, were reminders for me. To blame that man would be to travel down a dead-end road. To reveal my anger would be to shackle my voice. I struck up a conversation instead, and let the castaways, some of which I carried in a plastic bag, tell their stories. The man listened. I hope he heard. I hope he called his local municipality and told them to coordinate mowing with the seasons.

Calling city works may not be revolutionary. It is not sexy. But each phone call matters. If we all tell someone, and they tell someone, then the whole world will know.

Because I couldn't inspect the milkweed nor ignore the displaced caterpillars, I opted to leave the back-road bike route ten miles early. My heart needed a break from the slashed shoulder, and the highway's gravel edge was an escape.

At the Trans-Canada Highway, the distraction I sought was revealed. The highway had looked remote and adventurous on a map, but in reality it was burdened by a restless and unending line of traffic. Holding a straight course took center stage, and for the next few hundred miles, there was nothing more than a few inches of shoulder between traffic and me.

At lunchtime, just as a thick rain blasted from the sky, I found a country church with a perfect, protective overhang. I stood dry while I washed my hands with the steady stream of water pouring like a faucet from the edges of the roof.

As I began to fix lunch, I made a disturbing discovery. I no longer had four panniers. One, two, three. I counted again. After a minute of denial, I was forced to admit that one of my rear panniers, stuffed to the brim with my cookware, food, and bike tools, was gone. I ran through the last few hours, trying to imagine where it could be. I would have noticed had it fallen off as I rode. My only theory was that it had come off when I'd laid my bike down fifteen miles earlier.

Panicked, I started biking back, but minutes in I gave up on that idea. I couldn't afford adding thirty miles (half a day) to my schedule. Next I tried hitchhiking, but the traffic was too fast. As my efforts failed, I imagined my pannier picked up by a curious driver, smashed by passing semis, or tumbling down the shoulder and out of view. I walked up to a lonely house and knocked on the door.

The lady who answered saw how worried I was and led me into her house. Being hungry intensified my panic, as I imagined the effort it would take to replace everything, including my cookpot, which was my most prized possession. I knew it was just stuff, but I didn't want to lose my stuff in such a foolish, careless way. It weighed heavy.

Sensing my despair, the woman sympathized with my predicament. Her compassion made me feel silly and dramatic, and I started to cry. I was too old for this. I wanted to kick myself, but instead I let her take care of me. She made me a sandwich, got me some water, and then we took a drive. Indeed, fifteen miles back, where I had laid my bike down on the shoulder, sat my missing pannier, like an obedient dog waiting for my return. I thanked her profusely for her kindness to a stranger.

At the end of my riding day, I realized that the delay of my stumble had brought me to a tiny sliver of public land where I could camp on the shores of Lake Huron's North Channel.

Well, at least try and camp. Lake Huron was a quarter mile off the road, guarded by a jungle of trees, rocks, and mosquitoes. My craving for a quiet night on the shore of one of the Great Lakes motivated me to dive into the thick forest and see if I could get to the water. Pushing, pulling, dragging, yanking, sliding, hoisting, and lugging my bike through the morass of woods, I got closer and closer.

It was only after a tree branch whacked my face as a vine tripped me and a rogue stump gouged my shin that I second-guessed my bushwhack-with-bike decision. A small army of black flies hovered too close, and I translated their buzzing into a scornful What are you doing? Yet, as soon as I saw the turbulent water crashing against a horizon of craggy islands, I knew it had all been worth it. The forest muted the road noise. I saw no human footprints. The mossy rock world was mine to explore alongside adventurous tent caterpillars. My reflection rolled in ripples across the lake. The black flies buzzed and bit, like guard dogs, but I didn't mind. When the stars began to dance with the waves, I retreated to my tent. Through the mesh, the thwarted flies became a humming lullaby.

The flies, patient (and hungry), were waiting when I awoke. Since I didn't carry bug spray, I armed myself with raingear—a hot, plastic fortress against their attack. Bike packed, I batted and cursed at the flies as I retreated to the road.

Sweating and panicked by the buzz of so many wings, I jumped on my bike without ceremony. Two miles down the road, I stopped to shed my raingear. Those first few moments, as the fresh air touched my drenched skin and washed away my flustered escape, were more refreshing than a cold soda on a hot day.

Though the flies were thick at times, I found most of the creatures sharing the Trans-Canada Highway to be hospitable. I had grown most fond of the tent caterpillars who seemed determined, yet aimless. I stopped to help many cross the road, which sometimes took the form of simply flinging them out of danger. For many folks, they were pests and nothing more. How sad, to see only nuisance in such beauty. I studied the gold flecks and emerald green swirls that marked each hairy caterpillar.

By the time I made it to Sudbury, a city of lakes and redemption, I needed a break. Dave Lickley and Heather Jeramaz, friends of friends, had invited me to stay at their house, which sat like an island among vegetable gardens and aspens on a charming lake's edge. They were away on a weekend canoe trip, so I used the key they had hidden to let myself in. Their generosity and trust followed me inside.

In the kitchen I found a welcoming note instructing me to dig through their fully stocked fridge and eat to my heart's content. Needing no further cajoling, I served myself some leftover homemade stew. Full, and temporarily forgetting my overflowing to-do list, I permitted myself a few luxuries. I showered, did my laundry, and, with a cup of warm tea, I watched the building storm's rain finally fall. It was a school night, so I went to bed early.

Unlike in the United States, where summer break was in full force, Canadian schools were still in session. I woke up early to prepare for my first Canadian presentation. The students greeted me with enthusiasm, gave me their rapt attention, and rewarded me with both poignant questions and a ziplock bag filled with money from a school fundraiser. Those coins were worth more than what they could buy. They were tokens, reminding me that people believed in me and were cheering me on. Having so many kids honor me with their optimism, faith, and support made me want to live up to their vision of me.

After my day of presentations, I treated myself to a rest day, something both my mind and body needed. Dave and Heather returned home in time for me to meet them, and for them to fire up the stove for my first-ever sauna.

Sudbury, with a large Finnish population and plenty of cold lakes, was a sauna-loving community. Dave began by building a fire in their small sauna room. When the splashes of water on the fire sizzled, we filed in. I, the greenhorn, sat the farthest from the fire. We simmered in the humid heat as it turned from inviting to oppressive. When the appeal of the cool lake beyond was too tempting to ignore, we melted into its water. When the cold water overpowered our bodies' heat, it was time to repeat. Back into the soothing blanket of heat, then back into the lake.

Sudbury was the story of that lake, its neighboring lakes, and the forest that connected them. It was a cautionary tale of what can go wrong and what can go right. I heard Sudbury's story many times.

By the early 1900s, Sudbury had fallen victim to the pursuit of copper, nickel, and other minerals. The by-products of such mining were barren, blackened hillsides and soil erosion. Sudbury became a nearly unlivable wasteland; the dead lakes sat like sterile reminders of the consequences of human actions. Over the decades, Sudbury came to resemble the surface of the moon—so much so that in 1969, Apollo astronauts trained there. Unlike climate change or microplastics, the consequences were obvious, impossible-to-ignore reminders of overuse and imbalance.

Then Sudbury took a stand. Doctors, teachers, biologists, and artists fought death with their science, innovation, and passion. They balanced the acid in the lakes, planted seedlings, and watched their efforts succeed. Everyone's combined energy had helped recreate surrounding forest and made our swim in clean water possible. Another glimmer of hope.

I stopped to study a patch of milkweed at the edge of one of Sudbury's revived lakes. The story of the lake was one of second chances. For the monarchs and milkweed, it was a story of first chances. Had the forests of Sudbury not been cut, successional plants like milkweed would not have been given space to flourish. The milkweed had benefitted from the dramatic disturbance, as had the monarchs. Like a wandering traveler, nature took advantage of whatever it stumbled upon. The monarchs and I traveled the land, looking for opportunity.

East of Sudbury

MILES 4233–4593

I inched along the Trans-Canada Highway, joining the slow march of spread-out cyclists. Like all of Canada's cross-country traffic, cyclists were funneled onto the two-lane highway by the expansive, roadless forest beyond. I began to see several cyclists a day, a wild jump from the one I had seen in the previous 4000 miles. Still, the number was low enough that each passing held a moment of excitement. If we were on opposite sides of the highway, we would acknowledge our shared undertakings with a wave, nod, and/or the chimes of bike bells. If we passed on the same side, we would stop to exchange digests of our trips. Even covering the same section of road, our stories and experiences differed. Like any other gaggle of creatures headed the same direction, we shared much in common, but each of us was our own story. Sometimes our stories would meet, link, and transform the day.

Nick, Ryan, and Frank had caught up and then passed me just before a thunderstorm crackled down the highway. Turning on my blinking lights, I wished I could have kept up with them. Alone, I felt invisible in the slurry of rain. Cars rocketed by, like bullets I needed to dodge. At my first chance, I steered away from the menacing traffic and onto a slower, saner route twisting through the town of North Bay. That was when I spotted the three of them sheltered under a restaurant's awning. I parked my bike in a puddle of rain alongside theirs, and we all escaped inside to swap sagas from the road.

Nick and Ryan, two guys about my age, were cycling to New York City in a style completely opposite of mine. They lugged as little as possible on their bare-bones bikes. They were going for speed and thrived on cranking out uninterrupted miles. They slept in hotels, ate in restaurants, and didn't see the appeal of my approach (carrying everything and going slow). They had met Frank on his TransCanada trip several weeks earlier, and the trio had been meeting up each evening for hamburgers and to split the cost of hotel rooms. Even though I had been planning to bike for several hours more, their camaraderie and the cold, scary rain, gave me pause. As much as I needed the extra miles, hanging out with cyclists would be more memorable. I took them up on their offer to share a hotel room.

In the world of bicycle touring, it is common to meet and travel with a stranger for an hour, a day, a week, or even longer. Fellow travelers size each other up, and once there is a mutual sense of legitimacy and general goodwill, there are benefits to sharing resources. This is done not without hesitation, but with trust that outweighs the hesitation.

That was my rationale for sharing a room with three near strangers, much to the dismay of the woman working at the front desk of the hotel the next morning. As I left, she asked me which room number I'd been in. I told her the truth, which was that I had no idea. Next she asked me who paid for the room, and I replied, "Frank." I didn't know his last name. She looked at me with great contempt. I thought about explaining, but saw no reason. I smiled as I let myself out. I encountered many more cyclists in the days to come, but spent only a few minutes with all but one.

Jim was both a doctor and a preacher, and he caught up with me when I stopped for a photo shoot at Quebec's roadside welcome sign. Since his style matched mine, it was natural to bike into Quebec and the shadows of the setting sun together. "You going to camp?" I asked him. No explicit invitation, and no assumed response. "Yeah," he replied as we continued in tandem. When we passed a roadside park perfect for camping, there was no discussion. Our trips, at least for the night, would overlap.

Around the picnic table, we exchanged stories and contemplations. Jim, sixty-eight, told me he was traveling across the country because he was "preparing to get old." It reminded me of how, many years earlier on my first big tour, I had received advice from people to "do it while you are young." Though not wrong, I knew it was only partially true. I had met fifteen-year-olds and eighty-five-year-olds on remarkable adventures. The better adage may simply be to "do it while you can." Jim understood this. "No one ever says they wished they'd worked harder," he mused.

Though Jim and I were both self-contained and traveling at a similar pace, there were still plenty of differences. Jim had a hammock. I had a tent. Jim cooked a warm meal. I made a sandwich. Jim washed his dishes with soap and hot water. I used a piece of bread as a sponge. In the morning, Jim woke early, packed up in the rain, and left with the fortitude of someone who actually cleans his camping dishes. I stalled until an embarrassing 2 p.m.

Even following the same roads, no two travelers can ever have the same adventure.

My lazy start ended in cold, muggy disappointment. Instead of arriving at Margaret and Brian's cozy farmhouse that night, my delayed start forced me to bike until dark, set up my tent in the rain, and deal with another soggy night camping. I was forced to postpone the waiting warm shower, warm tea, and warm house until the next day.

When I arrived at Margaret and Brian's dairy farm, set on a dirt road (and looking like a rustic painting), the rain was still going strong. I arrived soaking wet, and stepping inside was heavenly. Showered, warm, and finally dry, I let the rain wash away the rest of the morning. By afternoon, the clouds had parted, and I walked through Margaret's garden, a freshly cleaned world saturated in sun. Each leaf shimmered, and the flowers were glad to be reunited with their shadows.

Touring the garden, I felt a powerful link between the monarchs and me: we were being cared for, fed, and housed by a community. I slurped a cone of homemade, double-chocolate ice cream made by Margaret with milk from their dairy. Mottled butterflies launched from their flowery snacks. Black beetles hunted among the leaves. The garden was a feast, and while some might have seen chaos and called for a mower, I was learning to see the world from the perspective of monarchs and their wild cousins. I saw potential in a jungle of green. I saw beauty in the less tamed and less curated. I saw perfection in the unruly. I saw Margaret feeding the migrating monarchs and their hungry caterpillars. The monarchs and I both needed people like Margaret for our journeys to come to fruition.

In return for the flower nectar, the butterflies repaid Margaret by shimmering in the sun like nature's stained glass. I thanked her for her generosity to both me and the monarchs with a small painting, a humble gesture to help convey my appreciation. Then I pointed my bike toward Ottawa.

It was looking like the capital city's Canada Day fireworks extravaganza was going to align with my arrival. I had no place to stay in Ottawa, nor even the remotest of plans. Still, I headed down the farm roads, full of skipper butterflies sitting like taxiing planes on a runway, toward the chaos of the city, the hype of Canada Day, and the infinite unfolding possibilities.

Before reaching Ottawa, I rode straight into a wave of blooming milkweed. Since Kansas, the common milkweed I'd seen along the roads had been in varying stages of growth. First it had been leafing sprouts, then flower buds. But like a satellite in space, enough ahead of the sun to never see a sunset, I had always been just ahead of the full bloom.

Now moving south for the first time on my trip, I charged into the wall of sweet-smelling flowers I had been running from. Milkweed blooms vary in color by species and can be white, yellow, green, pink, or orange, but it is the common milkweed's purple flowers that made me fall in love with the entire genus. Common but not *common. Asclepias syriaca* stretched down the road as if watching a parade go by, waving purple sparklers to celebrate all the days of summer. I got off my bike to revel in the pageantry. Taking long, deep breaths, I hoped to hold onto the delicious fragrance. It was easy for me to ask people to cultivate such a masterpiece. The plant earns its accolades.

The fireworks show in Ottawa didn't start until 11 p.m. (who starts a fireworks show at 11 p.m.?!). Like a drop of water, I bobbed through the claustrophobia-inducing crowd of Canadians and Canadian flags flooding the city. On a bridge spanning the Ottawa River, I basked in the view of the city. The water below separated Quebec and Ontario, and hundreds of boats, anchored in the current like cars at a drive-in movie, filled the channel. We all waited in hushed whispers. At the first bloom of color, the hush turned to cheers, and sound flashed overhead.

The first explosion lit the skyline. Each clap of light shimmered on the water below me, and the smoky finish blurred the silhouette of buildings. Without friends to join me in oohing and ahhing over each sizzle of light, I pondered the craziness of humans. We were together, thousands of us, to watch other people launch flying lights that rained color, smoke, and ash.

Twenty minutes later, Canada Day was over. The randomness made me smile, as did the absurdity of the plan forming in my mind. I was going to bike out of the city. *Now is as good a time as any*, I told myself as the crowd thinned. It was near midnight when I could finally pedal my bike once again.

It required a few wrong turns to find the bike path that could lead me through town. People in small clusters radiated out, headed toward homes where they would tuck themselves into bed. My plan was the same, though where I would unroll my bed remained a mystery. Unworried, I let the carefree feeling lead the way as I meandered through the people and the city. Neither asleep nor awake, I was seeing the city out of its routine. There was not a better time to navigate it.

The bike path wound out of town until my sense of direction, muddled at the best of times, was barely registering up from down. At least, thanks to the light reflected in the eyes of passing possums and raccoons, I knew I wasn't alone. Revisiting the city's many layers, by 2:30 a.m. I had crossed the outermost zone, where the countryside and faded metro mingled. The thrill of the night had worn away, and I stopped to investigate a church for possible camping. A parked car, a muddy yard, a weird vibe. There were too many warning signs, so I continued on.

At 3 a.m., I rode the perimeter of a rural school. Assessing my ability to be seen and projecting the morning traffic from each corner of the grounds, I ruled it out as well. A mile later, I wandered down a gravel road. It was closed to cars by well-placed boulders, and permitted only walkers and cyclists to pass. I scooted around, crossed a small creek barefoot, and arrived at a perfect, flat spot, fenced by wildflowers. Finally, seventy-four miles from Margaret and Brian's dairy, I was camped at a place impossibly perfect and impossible to have predicted.

Just as I had planned.

Welcoming "Weeds"

DAYS 113–123 / JULY 2–12

MILES 4593–4858

Barely into Vermont and heading east once again, I arrived at John and Nancy Hayden's fruit and berry farm: the Farm Between. The red barn and farmhouse, which signaled me off the road, were legacies of its dairy farm history. When I arrived at the top of their winding driveway, instead of cows, I saw that a seemingly endless sea of milkweed grew from the shores of the drive. Purple blooms capped each wave as the milkweed held steady in soil that had been worked for many years. Now, allowed to do what soil wants to do, it held the roots of its native plant community. Looking at more milkweed than I had ever seen before, I wondered if the Farm Between had been named for its location between two quaint towns or because here nature was permitted to thrive between the fruit trees and berry bushes.

Over the next few days, I walked around the farm's lavender canvas. I gulped the air left sweet by milkweed blossoms and listened to the language of insect wings testifying on behalf of nurtured ground. I met the young apple trees stretching in the sun, and tasted huckleberries and elderberry flowers snuggling with their wilder neighbors. I drizzled a stack of Nancy's homemade pancakes with a sampling of farm-fresh fruit syrups. I was another lucky migrant enjoying the bounty of their farm.

If the spread of wildflowers looked like a sea from afar, wading in, I found each milkweed to be an island of habitat teaming with creatures.

A googly-eyed spider peeked from a curtain of leaf. An angelic moth, satin-white, floated from purple bloom to purple bloom. A fiery legged insect, vivid from its recent molt, gained its bearings in the stubble of the milkweed's coarse white hairs. Upside down, a fifth instar monarch caterpillar straddled a milkweed's main stalk, and like a beaver, chewed at a leaf's stem.

I squatted down. The caterpillar was cloaked in the green cape of a chomped and almost defeated leaf. The leaf oozed its defensive white latex sap; the caterpillar was undeterred. It worked methodically, an architect confident in its plan. By chewing the stem, older caterpillars, like the one I watched, could divert the flow of sap away from the leaf they wished to eat. A pre-meal ritual akin to saying grace.

Young caterpillars employ strategies older caterpillars don't, because they are more at risk of getting mired in the milkweed's sticky, glue-like sap, and starving with their mouths gummed shut. First and second instar caterpillars defend against this by nibbling a horseshoe-shaped moat on a leaf's face. Cut, the milkweed oozes its latex sap, draining the small patch, and leaving an island of edible green for the stealthy caterpillar. It reminds me of my childhood friend who blotted the grease from her pizza before eating it.

Such defenses—both the milkweed's gooey sap and the monarchs' technique for avoiding its perils—are the results of a slow-motion, evolutionary battle between the two. None of us will live long enough to see the end game, but we can see what has already transpired. When the milkweed evolved toxic cardenolides, the monarchs evolved systems to sequester the poison and turn it into their own powerful defense mechanism. When the milkweed evolved sticky sap, the monarchs evolved behaviors to avoid it. The contest continues.

On a neighboring milkweed, I saw another battle playing out: a shiny island of shaved leaf and its caterpillar castaway enjoying the reward of its toils. The fuzz on the leaves of many milkweed species evolved to guard its green flesh, like hairy gates. The monarchs, unwilling to concede, evolved ways to bypass this armor. First and second instars bite off these hairs in order to access the leaf's flesh.

The milkweeds defend. The monarchs attack. It's an evolutionary arms race as old as their time here on Earth. If monarchs and milkweed can withstand human pressure, they could shift to something beyond our wildest imagination. Diversity needs only incentive, space, and time.

John and Nancy nurtured the farm's diversity, not only because they were committed to sharing the planet, but because they could put that diversity to work. The wildflowers brought in pollinators, protected the soil from sun and moisture loss, provided habitat for wildlife, returned nutrients to the soil, sequestered carbon, improved water quality, fed their horses, and gave balance to the farm. It was refreshing to see them put their trust in nature, and not outsource ecosystem functions to chemical fertilizers and pesticides.

Their farm was proof that we didn't have to choose between feeding ourselves and feeding the monarchs. We didn't have to strip the land bare. We could have the deliciousness grown on John and Nancy's farm—the full spread of fresh fruit, ciders, and fruit-syrup snow cones—*and* we could have monarchs. Moreover, our food and the monarchs' food could work together and deliver healthier farms, a healthier monarch migration, and a net of protection for other pollinators.

Protecting pollinators was John and Nancy's goal. Their farm was first and foremost a fourteen-acre pollinator sanctuary, a place to combat the worldwide pollinator declines. Whether you plant a garden for a monarch or a bee, you are creating an umbrella of protection for all the less-noticed species as well. Many gardens I visited aimed to protect monarchs, but by extension sheltered a full array of pollinators.

I watched a bee float from a tomato bloom to a milkweed flower. Once common, now every bee I saw was a survivor. From 2006 to 2013, commercial beekeepers saw the loss of 30 percent of their colonies each year. During the same period, wild bee populations declined an estimated 23 percent. Pesticides, habitat loss, climate change, parasites, and pathogens are all linked to this disturbing decline—a decline made more disturbing when we think about all the crops that rely on animals for pollination (including cacao for chocolate!).

John and Nancy knew how important it was to protect the bees. They relied on the bees to pollinate their livelihood. Their bee hotels—collections of hollow sticks in which the solitary bees could make their nests—and their colonial hives housing honeybees were scattered like welcome mats around the farm. Their grateful tenants worked overtime. Like door-to-door salesmen, the bees offered pollination in return for nectar.

John, a charismatic bug nerd (he attended bee camp) and beekeeper worked hard to keep the bees healthy and happy. When he spotted a swarm of honeybees in a vine ten feet off the ground, his joy was contagious. I quickly agreed to help him. His goal was to cut the vine and move the queen into one of his bee boxes.

John explained that the queen was in the center of the throng and the workers were going out on scouting expeditions in search of a suitable place for a new hive. His bee boxes were a perfect home, the bees just needed help finding it. Though during the swarming stage bees are extremely docile, we both donned beekeeper hats. Then, propping up a ladder and climbing face to face with the bees, John began to cut the vine and free the colony. My job was to stand under the swarm holding a large wooden tray, in which I was to catch any of the dislodged swarm that might fall from the main clump.

As John trimmed the bees free, a few began to fall. Then a lot. Many stayed down, but others flew dazed around us. I stayed as still as possible, while the bees determined my threat level. Not one of the hundreds of bees stung us as we placed the queen, her bubble of workers, and my tray of stragglers in the beehive. The others would find the queen settled and declare the hive home sweet home.

"Done," John declared, impressed. "Great job not freaking out!"

"Not freaking out?" I said, alarmed. "You told me there was no reason to freak out!"

An adventure is an exercise in trust. We must trust ourselves, the people we meet, and in this case, the gentle bees humming homecoming songs. We also must trust in the unknown, which will reveal itself in time. For a moment, my unknown was being covered in bees, on a milkweed-rich farm in Vermont. I had biked there, through the unknown, from Mexico. A different unknown beckoned me forward.

At the time, forward for me meant east, and east meant tackling a growing number of growing hills. Leaving the green refuge of John and Nancy's Farm Between I was met by steep, unrelenting climbs that slowed my progress and sped up time. All my muscles were forced to work in tandem as I slogged up roads with names like Devon's Hill, Farmstead Hill, and North East Hill. When the roads stalled on meadow ridges, I spun my pedals effortlessly, enjoying the view of green layers mingling with the clouds beyond. On the downhill sides, I thought of nothing but the pleasure of stretching my taxed body and the glory of the speeding flight. Reading the roads for potholes, I let gravity deliver me to the bottom of one hill while momentum carried me up the start of the next. Up and down, passing farms and forest, meandering east, with a pause at Jess's tiny cabin.

Jess Huyghebaert and I had become friends during our time at Humboldt State University. We both worked at the Campus Center for Appropriate Technology, a campus organization dedicated to organizing, developing, teaching, learning, and supporting sustainability. I was the groundskeeper, learning how to plant garlic, prune apple trees, nurture native plants, trellis the vines of kiwis, write newsletters, and train volunteers. I was surrounded by an inspiring team and together we ran from the normal, questioned everything, and never waited for someone else to solve a problem. Those lessons shaped me more than any class I took. They became my backbone.

The most important lesson I learned as a groundskeeper revealed itself to me like an epiphany. I was attempting to build a retaining wall when the burden of my insecurities broadsided me. I was a fraud. I had no idea what I was doing. I confessed to my friend Peter Lynch that I didn't deserve the job. Not long after, he showed me a photo of the groundskeepers that had come before me. In the photo, the team smiled with shovels and dirty jeans, blending into the garden in which they toiled. They looked at home, they looked confident.

"It looks like they know what they're doing, right?" he prompted. I had to agree, though this didn't make me feel better. Then Peter showed me a photo from a few weeks earlier. In it, my friends and I sported pickaxes and dirt-smudged smiles. "We look like we know what we're doing, too," he said.

The key was not being an expert before starting, the key was starting. Fake it till you make it. That moment gave me wings to start, and freed me from having to know that which I could only know from time and experience. That lesson has carried me thousands of miles. That lesson helped me write this book.

I can't remember if Jess was in that photo, but if she had been, she would have had the biggest smile and the dirtiest jeans. She was a hard-working pioneer spirit who was at home in any garden. Nine years later, she was an organic farmer, nurturing a growing harvest and cultivating the skills to one day run her own farm. She still laughed with ease, was wisely supportive, and comfortably nonjudgmental. Sure, both of us did things that would have made our more idyllic, younger selves cringe, but it seemed like we were still on our same, less-traveled paths.

At night Jess, her friends, and I surrounded a campfire made from scrap wood pallets and pine tree wedges, eating leftovers from a recently attended wedding. We were unpretentious kings sitting on tree stump benches and a car's old, extricated bench-seat throne. As fire and darkness mingled with our stories, I felt at home.

During the day, Jess worked at a local farm and I tagged along. Compared to the perennials at the Farm Between, the vegetables Jess grew required more weeded breathing room. Yet, on the fringes of the fields, a whole cast of nature was welcomed. Milkweed leapt, liberated—proof in each oblong leaf that monarch habitat and farms were not mutually exclusive. As a barefoot Jess towed a non-motorized seeder, planting a plot of future lettuce, I dove into the wild edges. There I found untamed neighborhoods full of "weeds" and bugs. My bike tour was also dedicated to them.

Along
the Atlantic

Rain and mountains were among the last hurdles separating me from a swim in the ocean. Leaving Vermont and entering New Hampshire was the range called the White Mountains. At the time, they seemed fittingly named. They wore an unapologetic white mask of colorless clouds. Only small pockets of visibility allowed for glimpses of the mountain's forest shell. The blue-tipped spruce and yellow-hued alder strained to retain their green under the diluted sun. Even the grey rock walls were not immune to the veil of fog. With only fleeting peeks of my surroundings, I was left to fill in the hidden views with my imagination.

Up and over the mountains, my freewheel sang on the descent and delivered me from the hazy cold to a warmer world. I stopped to photograph wild milkweed, pluck raspberries at a pick-your-own roadside farm, and visit with students at their school garden (it was an honor they showed up considering they were still on summer vacation). I crossed into Maine, reached the 5000-mile mark of my trip, and drew closer to the ocean.

If the rain and mountains had been hurdles, then the ocean was a fence, with waves guarding watery worlds. While I appreciated the power of its salty vastness, on past visits I had felt like I was pacing rather than walking its shores, like I was trying to find a gate or bridge that would allow escape.

Arriving in July to a sandy beach in Maine, 5022 miles from Mexico, I leaned my bike against a bench straddling runaway sand and encroaching pavement. Seeing my laden bike, beachcombers offered congratulatory smiles, assuming I was at the symbolic end of a coast-to-coast adventure. Boasting quick-dry underwear and ridiculous tan lines, I walked confidently to the water. I continued out, bobbing with the waves, until my bike shined small on the shore. For a moment, the ocean was not a fence. For a moment, I was given wings to fly through a sky of a different blue. It was similar to climbing a mountain in order to know the clouds or canoeing a river to know the current. Testing the edges, we find bridges to new worlds.

The ocean herded me and the monarchs south. For many coastal miles, a fortress of opulent estates sprouted like weeds. Still, I stole views of the mighty ocean and the monarchs using the sea as a backdrop. They were on the route of the Eastern Flyway, sandwiched between the Atlantic Ocean and the Appalachian Mountains. Pushed east by prevailing winds, and taking advantage of the thermal lift created by coastal currents and mountain ridges, the monarchs were pressing south, as if in a funnel. Unlike the larger Central Flyway, which spans the Midwest, the monarchs of the Eastern Flyway don't have a direct path to Mexico. This longer route disadvantages them, and they lag behind during the fall migration south (usually by about two weeks). The longer route, from the northeast rather than the north, presents an obvious question: *How?!* How do monarchs born in New York know to fly south before cutting southwest, and monarchs born in Minnesota know to fly south more directly? Scientists still don't have a complete answer for this. Perhaps, like sea turtles, they are true navigators, able to know where they are in relationship to where they are going. Perhaps it is something else.

One clue to consider involves preliminary studies of dislocated monarchs. In one example, monarchs from Kansas were taken to Georgia. When released upon arrival, they flew south in the same the direction as the monarchs in Kansas. However, when Kansas monarchs were held for nearly a week in a cage in Georgia, they seemed to recalibrate to the local area. When released, these monarchs flew in a more westerly direction, like the wild monarchs in the area.

Adding to the perplexity is the fact that monarchs born in the same area don't all go to the same colony in Mexico. Monarchs from New York, Minnesota, and Georgia are distributed among the multiple overwintering colonies. Broadly speaking, around 80 percent of recovered tags have been found at the El Rosario colony; 7–12 percent at Sierra Chincua and Cerro Pelón; and the rest at other colonies. Strikingly similar percentages are found when you zoom in to their tagging origins: 79.4 percent, 84.5 percent, and 79 percent of the recovered monarchs tagged in Cannon Falls, Minnesota; Lawrence, Kansas; and Iowa, respectively, were found at El Rosario. Similarly, when 2800 monarchs were tagged within four hours at the same location in Kansas, the tags were still recovered at three different overwintering colonies at previously noted proportions.

I rooted on each monarch doing its thing, despite my lack of understanding. The common milkweed, decorating our coastal path, cheered too. I headed south toward Ipswich, Massachusetts. There, I stopped for a break at Katie's house.

Katie Banks Hone owns a business and website called The Monarch Gardener. She has been gardening for monarchs for many years. After transforming her lawn into a native retreat, she began helping others do the same. She gives presentations to local groups and schools and consults with landowners wanting to grow butterfly gardens. When I arrived, her yard was in full bloom, and her two young daughters, already scientists, gave me the grand tour. Inside and out was a labyrinth of nature's many curiosities. Questions and answers bloomed, crept, flew, slithered, munched, dove, hid, waved, sang, beat, danced, and thrived.

Katie was well informed about monarch science, including the controversy around a milkweed species called tropical milkweed (*Asclepias curassavica*). It is a species not native to the United States, but has become a popular, widely available plant, easy to propagate and grow. However, it is also a species that has brought up concerns. The two main issues are specific to where it is grown.

In northern latitudes, tropical milkweed doesn't survive the winter, unlike the native species that die back to overwinter in dormancy.

Tropical milkweed must be replanted each spring, like other annual flowers. Because of this, in the fall, when the aboveground native milkweeds are dying back in preparation for winter, tropical milkweed remains lush and green. Along with lower temperatures and decreased sunlight, eating lower quality native milkweed in the fall triggers developing caterpillars to emerge from their chrysalides as nonreproductive migrants. Emerging science suggests caterpillars that feed on the lush, green leaves of tropical milkweed may be more likely to eclose as reproductive adults, tricked into thinking there is still time to mate and lay another generation of eggs. Instead of migrating, they may lay a generation of eggs that will not survive the winter.

In southern latitudes, the issues are different. Especially in the southeast, temperatures may not get low enough for tropical milkweed to die, and it has the potential to become localized. Monarchs that encounter this milkweed have been known to break sexual diapause and become reproductive and non-migratory. This may seem beneficial because it could create pockets of monarchs that don't depend on the forest in Mexico to survive. The issue, however, has to do with disease, especially the protozoan parasite, *Ophryocystis elektroscirrha*, or OE.

OE starts its life cycle as a spore. Spores transmitted to milkweeds are eaten by caterpillars and multiply inside the larvae. After metamorphosis, the spores are on the butterflies' scales. Infected adults can spread these spores when they mate, or to the next generation by visiting milkweed plants and laying eggs. The cycle repeats. Severely infected monarchs often fail to eclose or elcose too weak to fly. Mild infections result in shorter life expectancy and weaker flight.

The relationship between monarchs and OE is another example of complex biological equations. The monarchs that leave Mexico in the spring are relatively infection free. This is because those that had high levels of OE likely couldn't survive the migration. In places where monarchs are dense, small levels of OE can build quickly, especially when the same plants are used for multiple generations. Thus, the denser an area is with monarchs, the more likely those monarchs will be infected, especially as the season progresses (and the milkweed hasn't died back to reset its spore loads). It helps that the summer breeding area is so dispersed that there will be places where infection levels are lower. Monarchs from these places will be the ones that survive the winter and recolonize the range in the spring, at which point the milkweed, having died back, will re-emerge infection free.

If tropical milkweed grown in the southern United States can survive the winter, then it won't die back and reset its OE spore loads to zero. Instead, OE persists and infection levels continue to build. This is a concern, not only for the monarchs that become nonmigratory, but for the migrating population that returns in the spring and is met with highly OE-infected milkweeds.

To further complicate the issue, tropical milkweed has a higher concentration of cardenolide toxin than native milkweed. This makes tropical milkweed more appealing to egg-laying females, who want their caterpillar kids to have access to the most potent defenses. The toxins help lower OE infection levels in caterpillars, but don't eliminate it. So those caterpillars are still infected, yet they may be fit enough to fly farther, mate more, and as a consequence, better spread OE, exacerbating the problem.

Also worsening the problem may be climate change. Tropical milkweed seems to increase in toxicity when grown in warmer temperatures. Compared to the native swamp milkweed (*Asclepias incarnata*), tropical milkweed becomes more toxic when exposed to warmer temperatures, perhaps too toxic even for monarchs. In research, survivorship of monarchs reared on tropical milkweed grown in warmer temperatures decreased compared to monarchs reared on tropical milkweed grown at ambient temperatures. This information is still new, and not yet properly replicated, but the implications are well understood. Tweaking nature, by introducing plants or changing the climate, has unforeseen and complicated effects.

Katie takes all of this into consideration as she recommends plants for growing gardens. In 2017, she figured the benefits of growing tropical milkweed outweighed the costs. Since she was so far north, her tropical milkweed died each winter, so OE was not given a safe winter harbor. The extended caterpillar season seemed like a good thing, as it extended the opportunity for even more kids to meet and greet monarchs. As I write this book, however, her opinion has changed and she has stopped using tropical milkweed. "I have decided to err on the side of caution," Katie told me, "and I feel good about that decision."

At first, when Katie stopped growing tropical milkweed, some of her clients were disappointed. But everyone is adjusting, she says. There are so many wonderful native species to fill in the gap. It is easy for her to recommend her favorite native milkweed species, butterfly weed (*Asclepias tuberosa*). "It's the plant for everyone with a black thumb: plant it, water it a few times, then ignore it," she said.

Neither Katie nor I think that tropical milkweed is the monarchs' biggest threat. I would still rather see tropical milkweed than no milkweed at all, but I appreciate caution. I appreciate the acknowledgment that when we tweak nature, there are consequences. Such caution is not usually practiced in our culture. Short-term gain seems to be our only motive. How sane it seems to think beyond tomorrow and err on the side of caution.

After the tour of Katie's front and backyard gardens, we headed to her local beach—the result of cautious people who decided to hold back and not build upon every grain of sand. We walked to the ocean, across dunes spotted green with native plants, to the sand smoothed to perfection by long waves. Protecting such places, even if we can't yet grasp how saving them could save ourselves, is to err on the side of caution.

Leaving the wild ocean edge, I turned inland, attempting to skirt Boston and escape its traffic. In doing so, I rode past Walden Pond, near Concord, Massachusetts. In the mid-1800s, this pond inspired Henry David Thoreau's most famous work. It had been his refuge. Yet, in the time between the publication of *Walden* in 1854 and my arrival in 2017, inspiration had waned considerably. I passed a long line of idled traffic, waiting to enter the expansive parking lot. A sign warned visitors that the water was not safe to swim in. I didn't stop. Erosion, bacteria, non-native fish, climate-changed temperatures, and a surge in urine levels had overtaken the iconic landmark.

We all crave a pond to which we can escape, but we are running out of them. Now we must rely on slivers of public land that float like the ghosts of our planet's past, in the sprawl of hindsight, trying to teach us what was and is.

Past Walden Pond, I traversed Revolutionary War roads never meant for commuter traffic. The same rock walls that had once guided horse-drawn carriages now seemed suffocating as I fought to share the roads with autos. I didn't realize how tense I had become until I turned into the Assabet River National Wildlife Refuge, where I would make two presentations. My entire body relaxed; the relief was palpable. I coasted for a moment, shedding the burden of traffic.

The audience of children in the first presentation laughed at my jokes and were joyous when we explored the butterfly garden for monarchs. When they crowded into my tent (coming close to but not breaking the record of eighteen kindergarteners), I told them, "Only a mansion could fit so many people." When I lifted my empty tent with one finger I said, "I can lift a mansion with one finger, which means I am the strongest person in the world!" Everyone cheered. Then I called on a young girl to give it a try. She too was strong enough to lift a mansion.

The second group was adults, and they were much less impressed. My cheesy jokes didn't raise even a groan. I began to understand the pain comedians must feel when they bomb a set. Still, no one walked out.

My route was more bearable when I left the next day. A late start gave commuters plenty of time to get to work and lock themselves away before I even got started. Instead of traffic, I was met with a lush web of lonely miles. Massachusetts's backroads led me to a Rhode Island dirt road.

I could have crossed Rhode Island in a day, but the option of camping on a lake wrapped in a forest was too enticing to pass up. I eased myself into the water, my feet mingling with the twigs and detritus that lurked below. Neither warm nor cold, the water gathered me, and I hung between water and sky.

Dog paddling back to where my clothes waited, the feathered form of a barred owl caught my eye. I slipped from the water onto a boulder warm with sun, and watched it through the branches as it watched me. The wind mixed the tan and cream feathers running down its breast. Small brown feathers made circles like goggles around its dark, darting eyes. Connected by curiosity, we watched each other, and after a few moments it closed its eyes as if to declare me harmless.

When darkness absorbed us, I retired to my tent. Through the thin walls, beyond the conspiratorial whispers of shuffling leaves, a shrieking *caaaw!* cut like lighting through the blackness. There was a response, then a conversation began, as each caw was answered. *What in the world?*

Thanks to the power of cell service, I was able to call my friend Aaron Viducich, a bona fide bird nerd and adventure buddy, to get some answers. I told him about the extra-curious and unafraid barred owl, the lack of the more classic barred owl call of *hoooo cooks for youuuu, hoooo cooks for you allllll*, and the creepy calls and responses rippling through the night.

Without pause, he confirmed what I already knew: I didn't know much about birds. The shrieking calls were hungry, juvenile barred owls begging for food. The confident owl I had seen was likely a juvenile, inexperienced with humans, learning from me while I learned from him. I found it reassuring to know that owls must learn to be owls, must learn to sing their songs, hunt their food, and navigate their world. We were both teachers, and the forest was our classroom.

Thanks to the small size of the New England states, I was feeling fast and my progress felt real. I entered Rhode Island on a Friday, cruised through Connecticut on Saturday, and was checking out Long Island by Sunday. My sights were set on New York City.

I was enjoying Long Island's rows of flowering crops, quaint farm stands, and unassuming beach communities when a big black truck pulled up as close to me as it could. The driver rolled down the passenger window, and I braced for the inevitable. I had learned from biking thousands of miles that when people rolled down their windows, it was either to throw things, yell obscenities, or (true story) spray me with water. *Here we go,* I said to myself as I tightened my grip on the handlebars and kept my gaze forward.

The driver leaned over and thrust his hand out the window. I flinched. "Buy yourself something to drink," he hollered over the howl of his engine. From his fist flew a $10 bill. It was more than enough to buy a drink, considering my daily budget was around $10 a day. Stunned, I stared down at my first aerial donation, then waved a thank-you as he drove out of sight. That moment taught me two things. First, generous people existed and some of them drove really big trucks. Second, even after thousands of miles of biking, I could still be surprised.

On its western edge, Long Island erupted into Queens, which faded into Brooklyn. Each mile gave me experiences I could use to replace the pop culture–fueled images in my mind. I turned off Brooklyn's main thoroughfare and progressed from busy streets teeming with constant traffic, to calmer side roads lined with faded bike lanes, to even calmer residential streets. I let each turn carry me deeper into Brooklyn, until I jumped off my bike to walk the last stretch of meandering sidewalk. I slipped between the double doors of Erica Lim and Obe Ben Samuel's apartment building, and barely managed to stuff myself and my bike into the elevator for a ride to the twenty-second floor.

I had met Erica and her friend and coworker, Lihn Nguyen, four and a half months and 5406 miles earlier in Mexico. Along with Barb Hacking, a woman from Canada we'd also met, we had spent a day traveling to an outlying monarch colony, Piedra Herrada. Near the entrance we found the road flowing with monarchs in search of water. This spring phenomenon was at peak intensity, as the warm temperatures and dry slopes pulled the monarchs from the colony. We were lucky to have stumbled upon the colony, walked with it, and felt the current of the thirsty butterflies. Facing uphill, I had watched the streaming monarchs float toward me, diverting only centimeters before hitting my face. It had nearly been literal eye-to-eye contact.

Four and a half months later, Erica gave me a tour of her apartment and some tourist brochures. Off I went. The river of monarchs we'd experienced had dispersed, generations had lived and died, and I was now in a forest designed by architects. Though I saw little to remind me of the mighty oyamel firs waiting for the monarchs in the mountains of Mexico, I still examined the skies. The monarchs here would not be tourists; any that passed through were simply returning home.

I was the tourist. On a network of surprisingly pleasant bike routes, I tried to see what others saw—the possibilities and opportunities made possible by a jumble of traditions. There were songs of many worlds, endless shows, countless menus, and never-tired streets. Alone among people, I felt a loneliness not found in my wanderings through wilderness. Lost in the swirl of humanity, I longed for bears and snakes.

Having left my bike on Erica and Obe's balcony on my third day, I was on foot as I backtracked to their apartment. I was taking the subway, so I followed a set of stairs from the street level down. Confidently, I put my ticket in the turnstile and moved forward. *Whaap!* I was jolted by the force of the turnstile arm locking. It wouldn't budge. I backed up to buy another ticket as a stream of people filled in the space around me. The jolt on my second try was equally harsh. This time I watched the line of people for clues. What was I doing wrong? After my third unsuccessful try, I laughed. The gates ate my money, the arm blocked my entry, and the people kept moving around me.

On my fourth try, I somehow got it right; the turnstile slackened and I boarded a train. While the ride had turned unexpectedly expensive, I reveled in my success, thinking it would be an easy walk back to Erica's apartment. All I had to do was walk a few blocks, cross a lobby, take an elevator to the twenty-second floor, and unlock apartment 602. So I walked a few blocks, crossed a lobby, rode an elevator to the twenty-second floor, and tried to unlock the door—all in the apartment complex *next to* Erica's. Again I laughed. I was the first person to be biking the route of the monarch migration, but New York City stumped, trapped, and confused me.

Luckily, one didn't need to be a city slicker to discover the global food extravaganza New York City offered. Even when I was lost, I could travel from China to Japan, from Thailand to Ethiopia, and from Italy to the Caribbean within just a few miles. Though my budget rarely let me eat out, in New York City I splurged on one-dollar slices of pizza, Ethiopian food, and of course, ice cream in a fish.

"For dessert, let's get ice cream in a fish!" Our friend and Erica's coworker Lihn had joined Erica and me for lunch one day, and she made this declaration after we'd paid the bill. I looked at them both skeptically. "Ice cream in a fish?"

Having eaten still-living peanut beetles in Colombia, bull heart in Peru, and mayonnaise sandwiches in Missouri, I was less shy than most when it came to consuming whatever was served. Besides, I knew I liked ice cream and I didn't mind fish, so I followed Lihn to a small shop with a line that trailed out the door. Apparently ice cream in a fish was popular.

Just as I had decided chocolate ice cream and trout might be worth trying, I glanced through a window into the shop. The so-called fish in which the ice cream was served was merely a fish-shaped cone. I was relieved as I ordered my fish full of sesame seed–flavored ice cream and added Japan, where the craze had originated, to the list of places I needed to go.

As we walked off our Vietnamese lunch and Japanese dessert, Lihn excitedly led us through a door for foot messages. Even when she wasn't traveling, Lihn was a great traveler, able to find adventures in her own backyard. I surrendered to the experience, untying my shoes, even though I was self-conscious about having a stranger touch my feet. It was slightly painful, slightly soothing; I walked out happy to have my shoes back on.

Then it was time to say goodbye to New York City and find my way out of its dense embrace. I packed up my stuff and crossed the Brooklyn Bridge, again enjoying the city's serrated profile. Though I had seen monarchs a few days earlier, fluttering through Long Island's farm fields and at a gas station, and I saw tiny caterpillars and eggs clinging to milkweed under light posts and at a park in Queens, I didn't think I would see one in Manhattan.

Then I saw one in Manhattan.

He sailed with the wind, above tourists stretching to photograph the Statue of Liberty. Admiringly, I traced his path through the unaware crowd to a cluster of hyssop blooms. He humbly pollinated the purple stalks, flying from flower to flower, like a superhero navigating a garden skyline. I saw his orange wings as nature's white flag. *Peace,* he seemed to say as he connected two worlds. He wasn't asking for anything more than gardens sprinkled through our world, a planet shared with all.

I wondered what the monarchs saw, what my eyes were missing. I wondered if my path had crossed that of his parents or grandparents. I wondered if I would meet his kids or grandkids in Mexico, still a long way away.

Back Toward Canada

From New York, I didn't head directly south for Mexico. If it had been fall, I would have continued south with the Eastern Flyway monarchs, hugging the Atlantic Ocean and passing through Cape May, New Jersey. There the migration is funneled across the Delaware Bay, concentrating the monarchs and the enthusiasts that track them. Since my timing was too early, I veered northwest across upstate New York's mosaic of flowery fields, farms, and forests, toward Southern Ontario. There I hoped to meet a passionate team fighting for the monarchs.

It took just a day to feel very far from New York City. The expansive countryside held my attention. On a bike path I waded through a trailside prairie, finding monarch butterflies and tussock moths in their caterpillar forms, gorging contently on milkweeds. Like monarchs, milkweed tussock moths (*Euchaetes egle*) can store the milkweed's toxins and use them for their own defense. Unlike adult monarchs that advertise their toxicity with bright aposematic warning colors, adult tussock moths are drab, nocturnal creatures. Bright colors would be a worthless warning in the darkness, so instead, they have an organ that clicks and is detectable by bats. Aposematic sound. The bats learned long ago that the clicks are best avoided.

I watched the tussock moth caterpillars, each a mess of black hairs, orange lashes, and white whiskers. Unlike monarchs that distribute themselves among many plants, the tussock moths stick together for much of their caterpillar lives. Together they formed a shag carpet. The leaf on which they gorged was nearly a veined corpse.

I looked up only when a fellow cyclist paused from his ride to ask me what I was up to. His biking partner had continued on, so I kept my answers short and gave him a business card as an addendum. He thanked me and left to catch up with his friend. No red flags. I resumed my explorations in the weeds.

Several hours later, I got a call from "Mitch," the cyclist I had spoken to, inviting me to dinner. It turned out he lived about ten miles up the road, and with two hours of daylight left and sixty miles under my belt, it was the perfect distance. I accepted. I rode to his house and was greeted by his dogs and him returning from their evening walk. The grill was warming up, and we talked bikes (because he was a cyclist) and education (because he was a retired principal). There were still no warning signs. His grown daughter had joined us for dinner, more proof that he was legitimate and I could trust him. After she left, I felt guarded but safe saying yes to a soak in the hot tub. In spare clothes, I relaxed in the tub and everything seemed fine, if not a little random. Mistake number one.

Showered and dried, I continued talking with him as we watched baseball. I told him about New York City and how much I loved the connections woven from serendipity. From Mexico to a foot massage. Mitch told me he had trained as a reflexologist and knew how to give foot massages. He reached down, grabbed my foot, and started massaging it. I had less than zero desire for a foot massage from this man, but my training as a woman and my status as a guest made my refusal indirect, subtle, and, apparently, ineffective. "No thanks," I said smiling, "I'm good. No need for a foot massage." My body recoiled, but he didn't seem to notice. It was true that he was doing exactly what the professional masseuse in New York had done, so at least I knew he wasn't lying about the training. It was also true that every cell of my body was aware of my vulnerability as my mind ran through escape options. He was a principal, a cyclist, a dad, but still I prepared to fight. As he returned my foot to me, I felt deep relief—my second mistake.

Then he cupped my face in his hands and tried to kiss me.

I pulled away sharply, unable to make eye contact. I was in his house, my bike was unpacked, the sun had set. With my limited options, I managed a quick, "Goodnight," and ducked into the spare room, closing the door behind me. There I fought back a million bad-case scenarios. *How could I have been so stupid? What am I going to do?* I armed myself with my phone and pocketknife, put all my stuff in front of the door so that I would be warned if he tried to enter, and decided I felt protected enough to stay.

In the morning, things were civil. He pretended nothing had happened, and I stayed guarded, waiting for an apology that never came. I rushed through the morning, loaded up my bike, and once I was finally alone on the road, pedaled as fast as I could, trying to leave the night behind. Miles later, the icky feeling lingered. I felt like an idiot for having put myself in that situation. I felt confused about my responsibility to object while being a courteous guest. I felt angry that he had ignored my obvious discomfort and took no responsibility for righting a wrong. I felt used. Worst of all, my confidence at being able to read people and make good judgments, the core of bike touring, had vanished. I reminded myself I was fine, and did the only thing I could. I biked farther and farther away.

As if reassurance from the universe, six monarchs caught my attention, prancing between the delicate pink flowers of the thistles that lined the road. Soon a parade of butterflies, from drab to flamboyant yellow, joined the pageantry. I watched the flitting colors navigate the sharp thistle spines to steal pulls of nectar and absorbed myself in the tiger stripes and jeweled accents of the swallowtails. With a careful eye, I let the browns and creams of more humble wings guide me through their world—until Mitch was far away.

I rode until the day gave up, finishing a long climb before tucking myself into the woods. On a hillside, I found a space almost flat enough. Then I stuffed spare clothes and my empty pannier under the down-sloped side of my air mattress so I wouldn't tumble away. I was grateful to be alone in the woods, no creepy foot massages, no awkward navigation. The forest wrapped around me, and I could relax completely in its refuge. I slept deeply and prepared mentally for another homestay the next day. I wasn't sure I was ready.

There was no need to worry. Monarch fans Don and Bruce were the perfect people to stay with after my unnerving night with Mitch. I felt at ease among their flock of pets and wild visitors: three dogs, a tortoise, metamorphosing frogs, and dozens of monarch caterpillars. I felt safe as we followed the unpredictable path of a long, free-flowing conversation.

Don Bradford and Bruce England were artists, educators, and the perfect amount of interesting. On summer break, their mornings were devoted to caring for their critters, and their afternoons were spent in their garage-turned-art-studio, sculpting clay. As the dogs lounged under an umbrella in the front yard, and milkweed waved in the wind alongside their house, I joined them, forming clay monarchs on bikes.

Inspired by Don and Bruce's lifestyle, I also spent time during my visit envisioning a future in which I could pause long enough to know the space between seasons, long enough for the migrations to come to me. Traveling opened my eyes to my many possible futures and the endless array of lifestyles. Often people were stressed about the state of their homes upon inviting me in. I always felt like a doctor as I reassured them. "Don't worry," I would say, "I've seen it all." Having stayed in tiny trailers and mansions, in houses both too clean and too dirty, in locker rooms, museums, barns, classrooms, fire stations, and dance studios, I knew there was no right or wrong way. I knew I wanted to find a place where I could live in a small cabin, bike to city council meetings, welcome frogs to my porch with a pond, and know every tree that cast a nearby shadow. I added having an art studio to my vision as Don, Bruce, and I lost ourselves in the clay.

Yet, even though I could linger an extra day, I was still a wanderer. I was still on a bike tour, during which every hello was matched with a farewell. I said goodbye, healthier than when I had arrived, and I let the open road fill the void left by leaving. The rosy blooms of Joe Pye weed bent with the wind, the faded farmhouses shrank into uncultivated fields, the pyramids of hops rose green and smelly, and the road carried me until . . . *Whoa!* I slammed on my brakes for a double take.

It wasn't, as I had first thought, the tiniest of hummingbirds, but rather a moth in a hummingbird veil. Zipping gracefully in and out of the flowers—its tiny, window-like wings framed in rusty scales—the hummingbird moth sewed invisible patterns from petal to petal. The road was forgotten as the sun captured the flowers, the flowers captured the hummingbird moth, and the painting they created captured me. We were like Russian dolls stacked along the road, a train of focus from small to big.

Continuing into the heart of New York State, the riding was good, the monarchs were plentiful, and my spirits were high. Not even a memorable but gross bite into a maggot-covered piece of Swiss cheese during a snack one day could make me feel anything less than lucky. I rode the hills that divided the glacier-carved cat scratches of New York's Finger Lakes—a corrugated world of forests and water, decorated with sightings of fishing osprey, another startled skunk, and a gentle snake coiled in the sun. When the world was lit only by a slivered moon, I turned on my lights to find camping.

I found what I was looking for on a small tract of public land resigned to the trampling of cattle. The nearly dry mud was pockmarked by prints and resembled the surface of a dirty golf ball. By now, my routine was fast and efficient. I didn't pause until I had set up my tent, changed into my pajamas, blown up my air mattress, unstuffed my sleeping bag, donned my headlamp, made a pillow out of spare clothes, piled miscellaneous stuff in its predetermined corners, eaten a haphazard sandwich, and written a quick summary of the day's ride in my journal. Nighttime chores done, I paused to wonder at the stars in my neighborhood and let the miles slip from my muscles. Then it was time for work.

Unlike most people on bike tours, I occupied several hours most nights with office work. The blogs, videos, photos, handmade watercolor thank-you cards, route planning, itinerary logistics, and never-ending emails were all necessary to lend my voice to the monarchs' cause. My keyboard clicks matched the crickets' serenade and my screen glowed like a second moon. An unlikely office.

By Canandaigua Lake, my third of the five main Finger Lakes, the wind's warning and the brooding clouds finally delivered on their promise with a downpour. *A middle finger!* I mused as the cold rain dripped down my sleeves, each fold a funnel. My shoes filled with many tiny rivers. To stay warm and feel like I had some control over my comfort, I kept going, headed to Mexico via Canada.

After a soggy night in the soggy yard of a middle-of-nowhere church, I skipped the westernmost lakes and entered Letchworth State Park, a long, narrow park flanking the Genesee River's deep gorge and waterfalls. The first views of the canyon, the park's main attraction, distracted me from the betrayal I felt at seeing the park's manicured fields and road shoulders. The canyon's rotten rock arms pushed aside my darkening mood. They unfolded far below and far beyond the road. At the bottom of the canyon lay a daredevil river pouring over waterfall steps—a stairwell of agitated water. No room for manicured anything. Unseen creatures swam through the rapids below, and vultures, pigeons, and ravens rode in the river of air above.

Trapped by the limits of my body and the road, I stood still while I let my mind join the water curving through the canyon. I imagined the thrill of moving through a vein of untamed Earth. I felt the river's power. Humans could mow, spray, bulldoze, fill, drain, and kill, but in the end nature would win. I felt its certainty, its will. Tired but unshakable, the waters, rocks, plants, and animals would rise up. There would be casualties, but they would win.

Most of the water rushed downward, turning from blue to white as gravity mixed in air. But some defiant droplets splashed upward into a rainbow banner that proclaimed the nonconformists beautiful. While my eyes watched the water, and my body felt the misty edge, my ears filled with the noise of turbulent current. Loud and proud, the water was unapologetic.

In a limbo space at the park's edge, where I camped secretly, silently, respectfully, and illegally, I listened to the wandering water. Even as darkness fell, I knew the river would not stop running. Even after I'd set up my tent, and I was tucked inside, I knew the waterfalls still fell. Even now, far from that river, I find comfort in knowing its water still runs.

In my sleeping bag, in my tent, in the forest, in the darkness, with the heartbeat of the planet reassuring me, I thought of the monarchs. They, too, persist even when we cannot see them. They, too, will come like spring floods, cavort like rainbows in the mist, and leave with the cold for as long as we permit them their home. Migrants, wanderers, travelers, nomads— the monarchs make their home, for better or worse, wherever they find themselves. Yet they, like most creatures (including humans) find themselves most at home when a welcome is extended: the sweet nectar of a flower, the green of a leaf, the promise of water. A smile, a wave, a garden.

That night my neighbors were trees and I fell asleep on the welcome mat of their fallen needles. Canada waited for me seventy miles to the north. My plan was to spend the next night in this country, camping at the edge of Buffalo, New York. Then I would tackle the border crossing the following morning.

I knew my plan was folly by lunchtime the next day. Buffalo's suburban edge exploded without warning, demolishing any hope of easy camping. With no idea where to sleep, I pulled out my phone. Connecting churches and schools via Google Maps, I visited a dozen potential spots and rejected them all. The quiet corners of churches were filled with milling congregants. School fields, often abandoned in the summer, were packed with soccer players and cheering parents. Deeper into the city I went. Brick and pavement rose up and squeezed out my dwindling possibilities. The setting sun added its own threat. Each minute it crept lower and my options became fewer. I was starting to get desperate.

With light for one more scouting mission, I pulled into another church and felt a moment of relief. There was a perfect corner of trees growing protected alongside a fence. All I had to do was cross the nearly empty parking lot without attracting attention.

Too late.

Several people lingering near their cars watched me cautiously, and suddenly getting permission became my Hail Mary. I explained my trip and my predicament to several folks. They studied me and my bike before rendering a decision. No, I couldn't camp on the church's spacious lawn, not even the furthest reaches, not even if I was going to leave early the next morning, not even if it was dark and I was out of options. Something about insurance.

The stress of the moment boiled over in me. I knew it wasn't their problem. I knew I had no right to be mad, but I was desperate. "Jesus wouldn't have turned me down," I lobbed as I met their eyes and turned my bike back to the road. It was admittedly a low blow.

"Wait," one of the women said from her gurgling SUV. I paused, but had no patience for excuses. I wasn't going to apologize for making them feel guilty. The woman asked me a couple questions, and my replies were short. I didn't have time to entertain them.

"You're right," she finally said. "You can't stay here, but you can stay at my house. I live right over there." She signaled across the street. Now it was my turn to be a bit ashamed. My behavior had been unwarranted. Her invitation was gracious. "Thanks," I said. "That would be really nice."

Heather Schieder and her four kids—Mia, Lucie, Jude, and J. P.—drove across the street and I followed. They showed me their yard, and as I scanned it for camping spots I saw some milkweed growing in the margin between a fence and shed. Protected from the energetic kids and a bouncy dog, the milkweed grew tall and I made a beeline to it. Two tiny first instar monarch caterpillars munched away. I peeked at the undersides of the leaves. Two eggs sat patiently, their rambling lives nearly ready to begin. This yard was not just a yard, it was a necessary corner of the continent, on which the monarch migration relied.

Huddled together, we watched the hungry caterpillars, the likely grandchildren of monarchs I had seen in Mexico. Those caterpillars were the link connecting us, not just to each other, but to Mexico and Canada as well. In a whirl, the kids took off, rushing to other parts of their yard in search of more milkweed. While they surveyed, I told Heather about the belief of many that the souls of our deceased loved ones return as monarchs to visit us.

Heather's expression changed as this information settled in. Quietly, she told me that the following day was the day her baby girl, Maggie, would have been born had she survived to her birth. Perhaps a coincidence, it felt like more. We both looked back at the caterpillars, seeing them, perhaps, for the first time.

Science has taught me to embrace possibility. To never assume the arrogance of total understanding. Energy is neither created nor destroyed. Who am I to say that butterflies are only butterflies? Who am I to say that the flap of a monarch wing did not change the wind, which changed my course, which led me to a family looking for their daughter and sister? Who am I to say that when we die our energy is not given to monarchs so that we might have a turn navigating through this strange, beautiful, most impossible world on orange wings?

We left the tiny caterpillars for the night and headed inside for leftover pizza. Hours earlier, feeling deserted and indignant in a church parking lot, I could not have imagined such an evening.

I had every reason to get moving the next morning. I wanted time to explore Niagara Falls, then get to Ontario for an upcoming presentation. Still, the connection I'd found with Heather, her family, and the mystic monarch caterpillars in their yard tugged on me. I felt like I was exactly where I needed to be, and my time there was not yet finished. I felt like the family needed me, and I needed them. When Heather invited me to join them at a local lake the next day, I hesitated only briefly before agreeing. I could always bike faster and make up miles. Niagara Falls and Ontario could wait. After all, my trip was about more than just biking.

Decision made, we drove to the local swimming hole. It was half lake, half pond, with flowers, including milkweed, creeping up to the water. The overflow of sand speckled in the colors of congregating tadpoles. Basking amphibians welcomed us by diving frantically into the cool water. The kids bolted after them, and I cheered on both humans and frogs.

After the shore's stand of milkweed had been examined and startled toads had been caught, studied, and released, we set our sights on deeper waters. A heron tip-toeing the shore watched us as we wrangled a canoe into the water and ourselves into the canoe. Having paddled the entire

Missouri River a few years earlier, I admit the lake seemed especially small, but the canoe felt like home. In the stern, I felt the power of moving a boat purposefully, and the thrill of leading the kids on our own mini adventure. When we spotted a giant bullfrog sunning himself on a floating log a few feet from shore, we aimed for it. We approached it like a crocodile in a swamp. As J. P. leaned cautiously toward the frog, I leaned the opposite way for counterbalance. Slow . . . steady . . .

"Ahhhhh!" we all shrieked.

Two things surprised me. One, we were not in the water. Two, J. P. held a bull frog as big as his head in his hands. We were all giddy (except the frog) with astonishment, pride, and the pureness of being outside and learning by being characters in nature's story. I could have biked all day and likely not remembered one mile as clearly as I (even now) remember the surprise and wonder on everyone's faces. Minutes later the frog wriggled free, bounced around the canoe, and sprung for freedom under our lively protests.

The day wore on, and as quickly as I had arrived, it was time to bid farewell, a traveler's bittersweet baggage. Two days earlier we had been strangers, and now it hurt to say goodbye. Goodbye for now. I rode away, the last forty-eight hours swimming around me, waiting for the moment when I could organize them in my mind. Just as I was settling back into the rhythm of the road, a few miles from Heather's house, the familiar gurgling of her SUV caught my attention. She signaled me to a side road. As she parked, the kids tumbled out with tears in their eyes. A few last hugs before I headed to Canada.

Canada, Take Two

DAYS 153–160 / AUGUST 11–18

MILES 6005–6188

I was ushered into Canada this time by a bridge under construction and a near-interrogation from an angry border agent.

"Gun?" she asked with bored contempt.

"No," I said, bemused. The idea of carrying a gun on a bike tour was unfathomable to me. Never once, even when my neck hairs prickled from potential danger, had I wished for a gun. *I would rather bring a piano on a bike tour* I thought, but didn't say.

"Knives?" she continued viciously.

"No."

"Mace?"

"Nope."

"Then how," she deadpanned with a dramatic pause, "could you possibly defend yourself?" The words themselves felt like a threat.

"I run fast," I said through a playful smile. Unimpressed, she ran her eyes over my beat-up bike and my faded, button-down, thrift-store shirt. Finally, she shot me a cold stare that told me I was not to joke.

I was only sort of joking. Weapons wouldn't have made me safer and would have changed the trust equation I relied on. Bike touring is dependent on awareness, common sense, rationality, luck, and—most important—trust. Would you invite an armed traveler who didn't trust you to stay in your guest room? Never mind going to schools.

I didn't offer her my explanation for being unarmed. Instead I let her steady onslaught of questions unnerve me. Her icy demeanor quickened my pulse. The bridge, known as the Peace Bridge, seemed poorly named.

When the questions ended and permission was granted, I entered Canada and knew I was one of the lucky ones. For many, my trip was an impossibility, not because they were less brave, less adventurous, less athletic, less anything, but simply because they were born in a different country and their passport couldn't unlock doors. My border dramas were entertaining stories, not unresolvable dead-ends.

I looked up, hoping to see a monarch. I wanted to watch one moving freely between two countries. I wanted to vicariously fly across the river and be reminded that it is a shared link, not a boundary, fence, or limit. Rivers link mountain to ocean, city to town, sky to Earth. Scanning the skies, I saw only fluffy white clouds. One shaped like an elephant eating an ice cream cone moved from the United States to Canada. *That'll do*, I mused.

"No, No, NO!" I was yanked out of my contemplations by the frantic arm waving of another border agent. Reading his lips, his body, and his irritated expression, I saw where I couldn't go, but not where I was supposed to go. I wandered, frustrated, until a local cyclist pointed me to a pedestrian ramp, which took me to the calm of Canada. On a quiet road paralleling the churning Niagara River, I cycled alongside its waters.

In the next few miles, that water would free fall more than 160 feet. Over the next 500 miles, from Canada's New York border to Canada's Michigan border, I would navigate a jam-packed schedule. I sketched a calendar in my journal, circling the confirmed dates of each presentation. It was August 11 and 13, 14, 16, 17, 19, 22, and 23 were circled. I was about to plunge into my own rapids—talking, meeting, greeting, presenting, and learning. For now, though, I followed the long braid of water to Horseshoe Falls.

Prior to my arrival, I had imagined the falls to be prisoners, trapped by development. I envisioned houses, malls, and neon lights holding captive the broken views of waterfalls reaching down and hotel skyscrapers reaching up. But as I walked my bike toward the rumble of river and the halo of mist, I was pleasantly surprised. Sure, there were buildings, but the river felt spared, and I paused to watch as it flowed into the white world and over the edge. Just shy of the drop, in the calm eddies where fish huddled, frenzied terns pierced the water's surface and emerged with lunch. The river waited in an orderly fashion to tumble over the cliff, and the tourists waited in an orderly fashion to witness it.

The crashing water's roar drowned out virtually all other sounds. It replaced small talk and traffic noises with a great, whooshing soundtrack. The water launching from the top was a band of Ozark river–turquoise, a hue I have never been able to recreate with paint. As it fell, a gradient of blues and greens succumbed to white, and only passing cormorants and gulls broke the blank canvas. I listened and watched and imagined the swirl of liquids and boulders winning and losing in their slow-motion battle below. It was our planet at its best: humble yet mighty, merciful yet unwavering.

A monarch hovered high above the mist. We continued into Canada together.

Like the powerful waterfall I had witnessed, Southern Ontario's team of passionate and dedicated monarch enthusiasts is carving a better future across their land. Their drive is an unstoppable current. Visiting and meeting many of these activists, I felt like I was on a boat steered not by me but by the flow of their work. Though I had not originally planned to go through Southern Ontario, their enthusiasm made it clear, even from afar, that a route change was necessary. The several hundred miles it added was worth it.

After making presentations at Royal Botanical Gardens near Burlington, Ontario, and Greenway Garden Centre near Breslau, Ontario, I pedaled into Stratford and arrived at Barb Hacking's house. A retired schoolteacher, Barb had bonded with the monarchs after raising caterpillars, season after season, in her classroom. Each year, as monarchs metamorphosed under the noses of curious students, she watched her students metamorphose, too. The interaction of species created a progression: foreignness to familiarity, indifference to reverence.

While it was my first visit to Barb's house, Barb was not a stranger. I had met her in Mexico before the start of my trip. She, Erica, Lihn, and I had spent a day at Piedra Herrada colony, and Barb and Darlene, another spirited monarch steward, had waved me goodbye at the beginning of my journey. They had connected winter to summer by airplane, and now, 6137 miles and five months later, I was connecting the dots by bike. Their familiar faces, like Erica's and Lihn's in New York, were welcome breaks in the constant newness I was experiencing on the trip.

Now retired, Barb had found new ways of sharing the transformation. She raised monarchs in her dining room, delivering nearly full-grown caterpillars—along with instructions on caring for them—to her community. Barb would then explain the forthcoming process to the recipients, including the nomenclature. A chrysalis is a butterfly pupa. Cocoons, on the other hand, have a silk protective covering, and thus refer only to the pupae of most moths. The beauty is in the details.

On a tour of town, we checked in on her monarch babies and the folks tending them. Each person recounted stories, with awe and surprise, of witnessing their banded larva transform into emerald-green chrysalides. They studied the gold and black specks of the chrysalides, waiting to catch the shapeshifting creatures as they moved to the next stage.

Nature, shy in moments of vulnerability, rarely invites people to watch such transformations. Chrysalides can't run or fight, so they rely on camouflage to survive the ten- to fourteen-day transition. Most move off of their milkweed nurseries before pupating, some more than thirty feet, making them even harder to find—a plump green needle in a wild, weedy haystack. I had spent hundreds of hours looking for monarchs and had never found a chrysalis in the wild (though I did find one on the underside of a lawn chair in a backyard with milkweed). Like most people, I had to rely on temporarily domesticated monarchs, through glass and netting, to watch the feat of metamorphosis transpire.

Peering into one of Barb's small enclosures, I watched a chrysalis hang patiently, full of delicate secrets. If I watched long enough, the pigmentation of its scales would develop, and the green pupa would graduate to its adult orange, black, and white color. The thin, transparent coverings of its chrysalis would split and a scrunched and sloppy monarch would eclose.

The newly emerged adult would grasp its empty chrysalis, preventing a free fall, to hang suspended.

Once a monarch has eclosed, it is still not ready for flight. Eclosed monarchs hang suspended, their overly plump abdomens framed by small, limp, curtain-like wings. Before flying, they pump hemolymph (analogous to insect blood) from their swollen abdomens into their deflated wings. The typically green fluid, like liquid scaffolding, must harden before the monarchs can stretch their wings. After three or four hours, they are ready to fly.

The primordial cells that become each butterfly's wings—as well as its legs, eyes, and other adult structures—are present in the first instar larva. Watching a monarch stretch her elegant wings for the first time, still not ready to fly, I understand but can barely believe.

Barb and her farmed-out caterpillars were invitations to everyone caring for them—to both witness some of nature's most intimate moments, and to join the network of monarch stewards. She was connecting her community, creating a network on which the yearly wave of monarchs could depend.

"Do you like peanut butter?" Barb asked. Between the pizza and butterfly-shaped waffles, I could hardly believe we were still talking about food.

"Sure," I said. I mean, I didn't *not* like peanut butter.

Soon I would learn that I was being observed and notes were being taken during my time in Southern Ontario. I was being studied. But like a game of telephone, my original responses were subtly changing as they were retold. Statements such as "I like peanut butter" were traveling through the monarch advocates' network and becoming "I *love* peanut butter and can't live without it!!"

As a non-picky eater (though vegetarian when I can be), I ate just about everything. As a traveler, I believed accepting invitations to meals was as important as covering miles. Food that was shared linked me to place, connected me to a spirit of generosity, and allowed me the opportunity to give back, with gratitude, a place in my adventure. If Barb had served me turnips and American cheese, instead of food I actually liked, I still would have eaten every bite, asked for seconds, and declared, without lying, that it was delicious—not because of the taste, but because it was a symbol of kindness. And so it came to be that my traverse across Ontario was being fueled by (among other foods) peanut butter and ice cream.

By the time I stayed with Barb, I had been weaving stories of ice cream—and my enthusiasm for it—into my presentations. First there had been the man in Mexico, who surprised me with strawberry ice cream from a cooler on his motorcycle; then Margaret, who had given me chocolate ice cream made with milk from her dairy cows in Canada; and most recently, Lihn, who treated me to ice cream in a fish in New York. I added the last story, when, after a presentation on the East Coast, a young girl had come up to me, nearly in tears, expressing her horror that no one had offered me ice cream in the United States. I assured her that I had enjoyed so many scoops in the US that there simply wasn't time to talk about them all.

The more stories of ice cream I told, the more stories there were to tell. After a presentation in Stratford, Ontario, while I was chatting with the audience, a cone filled with swirls of peach soft serve appeared miraculously in front of me. Unbeknownst to me, people had raced out after my presentation ended, in search of ice cream for me. Such conviction, the same they were using to rally for the monarchs, was astounding. When I left Barb's, it was as much to work up an appetite for my next scoop at Bruce Parker's house in London, Ontario, as it was to continue the migration.

Bruce, a cyclist and monarch guy, knew exactly what I needed: a shower, food, internet, and space to breathe. He understood that while staying at someone's house was rewarding, it could be more exhausting than long days biking and quiet nights camping. We talked with ease, and I tested out some thoughts about raising and tagging monarchs that had been forming in my mind.

For a long time, I'd had reservations about rearing caterpillars. I was hesitant to believe people were "saving" monarchs by removing them from nature. Swapping the tried and true natural systems for artificial sun, disrupted temperatures, and atypical light patterns made me nervous. I believed in evolution. In the wild, the monarchs most capable of surviving and passing on the most suitable DNA were sorted out by weather, predators, and disease. Furthermore, caterpillars, especially those raised in high densities, were often exposed to high rates of parasites and diseases, like OE.

During my trip, however, my views had been changing. At Barb's, I had seen that most of the caterpillar caretakers I met were diligent. They disinfected their equipment regularly, didn't keep large amounts of monarchs together, and never shipped them any distance. Moreover, I was seeing positive outcomes of captive-raised caterpillars. I began to understand that these caterpillars, slipped from their ecological reality, were important sacrificial guides and teachers. Glimpses into their world were more important than confronting the pressures of ecology. Caterpillars could be bridges to wilder worlds.

Bruce understood this. Like many who raised caterpillars, he showcased the monarchs he reared so that his community could better interact with the monarchs' story.

We can bring the monarchs to us, but we must also bring ourselves to the monarchs.

Even as my opinion on raising monarchs shifted, there were details that made my skepticism simmer. Wild creatures have evolved in wild homes. Heated houses, lightbulbs, and hand-picked milkweed all likely distort monarch development. Still, I have come to believe that altering developmental variables is justified when we collect the most vulnerable caterpillars and then give them platforms to be teachers. They can help people learn that each time we restore a plot of habitat, we are rearing monarchs and their wild neighbors many times over.

The Crazy Monarch People, as I affectionately called the network of passionate monarch enthusiasts across Southern Ontario, go beyond rearing. Starting in the fall, these citizen scientists are part of the vast volunteer network that tags reared and wild-caught monarchs. Where tagged monarchs are found depends both on where the monarchs go and if the people they pass are looking. In Mexico, the guides at the sanctuaries are on the watch. Each found tag can be traded for a $5 (100 pesos) finder's fee provided by the Monarch Watch representatives that visit the sanctuaries each winter. Any found tags are coveted. Delicately, the lucky finders fold the tags into pieces of paper that are stored in wallets and purses until an exchange can be made. Again, I must confess to initial skepticism. I worried that even the minimal weight of a sticker could be a hindrance to a small, flying migrant. But such worry is likely unwarranted. If it was too much weight, tagged monarchs wouldn't be showing up in Mexico. I was also concerned about how those tagged monarchs are found. Though I believe most tags are collected ethically, I heard stories in Mexico of people shaking monarchs from branches and trampling fallen butterflies at night in search of tags. The question to me was not whether this was happening, but rather, did the benefits of tagging outweigh the sacrifices? I decided the answer was yes. The better we can understand, the better we can protect. Many monarchs have fallen victim to our pursuit of understanding, but they have left a legacy of knowledge. Perhaps the best we can do is honor the casualties and remember them as scientific pioneers—just as we remember the man who first tagged monarchs.

In 1937, Fred Urquhart, a biologist living in the Toronto area, began tagging monarch butterflies in an effort to understand where the monarchs disappeared to. Just finding a lightweight sticker that didn't wash off in the rain took many years. Eventually he landed on a version similar to a price tag. In 1952, through a magazine article, Fred began seeking volunteers to help with his tagging efforts. At first only twelve people responded, but by 1971 there were 600 volunteers. In 1972, Urquhart's wife and fellow monarch researcher, Norah, placed an ad in a Mexican newspaper, looking for volunteers farther south. Reflecting on this, I see those invitations, for ordinary people to be part of extraordinary science, as one of the most important breakthroughs in monarch research. By including volunteers in the mission, the tagging efforts advanced science and created a culture united around the monarch that still exists today.

I've met several people who tagged monarchs with Urquhart, including John Powers and Don Davis. Don has been tagging monarchs since 1967. On September 10, 1988, he tagged a male monarch near Brighton, Ontario. An estimated 2880 miles and nearly seven months later, that monarch was recovered in Austin, Texas, on April 8, 1989. The tag earned Don and the monarch the Guinness World Record for the longest migration by a tagged butterfly.

John Powers (who died in 2019) was perhaps one of the biggest personalities in the monarch world. The day I met him, he wore a Hawaiianesque monarch-printed shirt. With great ceremony, he presented me with a copy of the August, 1976, issue of *National Geographic*. On the cover was Catalina Aguado and millions of monarch butterflies. Catalina had married a man named Kenneth Brugger two years into his search for monarchs in rural Mexico. Brugger had gone looking for the monarchs after reading Urquhart's ad for tagging volunteers. Kenneth and Catalina traveled by foot, motorcycle, horseback, jeep, and motorhome, slowly finding clues—dead monarchs on roads, swarms in the sky, local knowledge—to home in on the location of the overwintering monarchs. On January 9, 1975, Kenneth called Urquhart with good news. "We have found them!" Kenneth reported excitedly. "Millions of monarchs, in evergreens beside a mountain clearing." They had found the colony now known as Cerro Pelón.

A year later, Fred and Norah Urquhart trekked up the mountain in Mexico to see the phenomenon with their own eyes. In the *National Geographic* article, Fred describes the heart-pounding walk with a morbid sense of humor: "Norah and I are no longer young . . . Suppose the strain proved to be too much? It would be the ultimate irony to have come this far and then never witness what we'd waited so long to see!"

Luckily, their hearts held out and the scientists were rewarded by a storm of wings; their life's work had been realized. From their backyard in Ontario to a Mexican peak, the connection had been forged by forty years of effort, several hundred thousand tagged monarchs, and a team of volunteers tagging and recovering butterflies. They had known where the monarchs spent their summers, and now they knew where the monarchs spent their winters. Understanding the bigger picture put the onus on every North American to protect the monarchs.

I thought about those forty years. I wondered if I had the strength to dedicate forty years to something.

The story of the Urquharts' 1976 visit to Mexico continues with poetic surprise. While the Urquharts were at the colony, a three-inch-thick pine branch broke under the weight of so many wings. Fred sifted through the spilled monarchs, and to his amazement, found one adorned with one of his white tags. A science teacher named Jim Gilbert and his class had tagged the monarch in a field of goldenrod in Chaska, Minnesota. It was a gift from nature, validation from Mother Earth.

Word slowly got out about the discovery of the monarchs' overwintering grounds, even though the Urquharts refused to disclose the location to anyone—including renowned monarch researchers such as Lincoln Brower. Though the map in the *National Geographic* article placed the monarchs several hundred miles north of their true location, there were other clues for Brower and another researcher, Bill Calvert, to go on. They knew the monarchs were 10,000 feet above sea level, in the state of Michoacán. This information landed them in the town of Angangueo, where the mayor, surprised by such interest in a butterfly, helped arrange their passage to the colony now known as Sierra Chincua.

I have been in that same Mexican forest, but can only imagine the joy that filled the scientists. Decades of hard work, dedication, and passion culminated in the hum of billions of monarchs. More monarchs than I have ever seen. More monarchs than the world has seen since humans slowly began erasing their waltzing spectacle. After the winter of 1976, when the western world collided with the colony, the monarchs' refuge began to erode. Knowing enables us to protect, but knowing also enables us to destroy.

I used to wonder how the monarchs would have fared if we had simply left them alone. I've decided that if we hadn't made the discoveries we did, they would have disappeared from our skies with a pain so quiet that only the wind would have wept. Knowledge is the only sword that can cut through harm being done in ignorance. I look to Urquhart and all who have followed his pioneering footsteps with gratitude and admiration.

Bruce Parker was continuing Urquhart's mission. He had begun tagging monarchs in 1998. Though not a biologist by training, Bruce had become a scientist through experience. His efforts and what he'd learned through observation had sculpted him into one. In notebooks, he kept records of the wing length and weight of each monarch he tagged, seeking answers by collecting facts. He was tuned in to the monarchs, and I was inspired by his enterprising methods.

There is contradictory information on whether monarchs born in the fall are morphologically different from monarchs born during the spring and summer. Early research suggested that as daylight decreases, larval development slows, and the result is adults eclosing both heavier and with larger wings. This made sense, as larger wings would be of great benefit to the long-distance travelers that rely on gliding to migrate efficiently.

Monarchs have two forms of flight: flapping and gliding. Gliding is the more efficient of the two, and is used extensively by migrating monarchs. They gain elevation in thermal currents (like many birds do), then glide in their desired direction. Their glide ratio is typically between 3:1 and 4:1, or for every three to four feet forward, they lose one foot of elevation. When the wind is favorable, however, they can flap every so often and still maintain their altitude. Larger wings have greater surface area, which increases lift and efficiency. Though it is not clear whether the heavier, larger-winged migrating monarchs born in the fall are born that way in response to the task at hand, it is clear that wing size (and shape) do effect efficiency and thus migratory success.

A more recent study of monarchs reared in summer and fall conditions found that while fall conditions did slow development, adult size did not correlate with those conditions. While bigger monarchs are better suited for migration, they might not be bigger *because* they have to migrate. Rather, monarchs eclose in a range of sizes every generation, a result of variations in the milkweed they ate as caterpillars. All sizes may start the migration, but the largest winged monarchs are the most capable of surviving the journey. The pressure to have larger, more efficient wings increases for butterflies born farther north (because they have to fly farther), which could explain why the monarchs born at higher latitudes that reach Mexico are larger than the monarchs born at lower latitudes. Only the longest-winged monarchs can survive the longest migration paths, to arrive in Mexico and be measured.

While morphological differences are less understood, the physiological differences are clearer between monarchs of the overwintering generation and the reproductive generations born during the spring and summer. Overwintering monarchs eclose in sexual diapause, likely in response to poorer-quality milkweed, lower fall temperatures, and decreasing daylight. Unlike the reproductive generations that begin mating four to five days after eclosing, overwintering adults lack the necessary hormone responsible for sexual maturity. This delay—which lasts until warm spring temperatures signal reproductive organs to mature just as milkweed in Texas is beginning to re-emerge—is what allows overwintering monarchs to postpone breeding all winter. It also allows them to wait in Mexico for the milkweed in Texas to re-emerge from winter dormancy.

A delay in sexual maturity is also what allows overwintering monarchs to outlive winter. While reproductive adults live two to six weeks, the overwintering generation can live six to nine *months*. That is enough time to fly to Mexico, endure winter, then return north in the spring to emerging milkweed. As if on pause, overwintering monarchs can wait out winter like prepubescent tweens.

In Mexico, the overwintering generation is called *generación Matusalén*, or the Methuselah generation in English, after the oldest recorded human in the Bible. Yet, we do not need to use religion to explain such a remarkable feat. Sure, Methuselah supposedly defied time and lived to be 969, but he didn't fly to Mexico without ever having been there before. He didn't settle in the trees of his ancestors, to wait out the winter, and return again in the spring. No, only the overwintering generation of monarch butterflies can do that. In the United States and Canada, the overwintering generation is known as the super-generation.

First we learned where they went. Then how. What will we learn next? Will the migration remain intact, so that more of our questions can be answered?

Monarchs are one of the most studied animals in North America, and there are still many questions to be answered. Tagging efforts continue, and the link between us and the monarchs strengthens. The tag is a potential answer, a promise, a declaration of love. So was rearing, and gardening, and rallying as monarch stewards pushed forward. So were the miles I pedaled.

Fence Erie

DAYS 161-166 / AUGUST 19-24

MILES 6188-6358

I knew I was getting close to Lake Erie when one monarch became a kaleidoscope of many (a group of butterflies is officially known as a kaleidoscope). The lake acted as a wide, wet wall that halted the monarchs' southern progress, and shooed each toward the lake's narrowest crossing. I followed them, as they gained concentration among the milkweed. The monarchs' blitz was encouraging.

While their numbers had plummeted in the recent past and I was getting only a taste of what had once been, there were still reasons to celebrate. There was, after all, something left to chase. There was still possibility wrapped in eggs glued to roadside milkweed, the bravery of a caterpillar alone in its universe of leaves, and the beauty of a granddaughter reflecting tiny suns in her scales while she probed showy flowers for nectar. I devoured each sighting with all my senses.

The surge of milkweed brought me joy and joy's dark opposite. A desire to protect each smooth leaf and the neighborhood of life it hosted was both overwhelming and impossible. The plants sat so vulnerable in ditches, the caterpillars never eating fast enough to keep the milkweed small and inconspicuous. It seemed inevitable that the plants would draw the attention of landowners who were oblivious to the architecture of life, and the monarchs' habitat would succumb to mowing. It made my heart hurt. I felt weak for leaving each plant defenseless. I felt a bitter anger, a poison that reached my soul, as people waved from their riding lawn mowers, blind to their destruction.

Darlene Burgess met me on the road with her bicycle. Darlene was a pivotal player in monarch conservation and being in her presence distracted me from my downward spiral. Together, we cycled the remaining few miles to her house (just north of Canada's southernmost point), on roads lined with milkweed that bent under the weight of purple petals and orange wings. She understood my pain, and helped remedy it with stories of her work. "Know what you mow," she said, speaking about a campaign she was initiating to question the green grass delusion.

We are told that manicured lawns are beautiful, that we must control nature in order to live with it, but that is a lie. Beauty is the give and take between plants and animals. Beauty is milkweed, ripe with exploding purple booms, feeding the shaggy maned tussock moths and bees and monarchs. How can we possibly judge so much life as unworthy?

Darlene's "Know What You Mow" campaign was just one cog in her monarch-guarding machine. She reared monarchs, stocked the Point Pelee National Park visitor center with caterpillars for tourists to meet, was an invaluable resource to new volunteers, tagged fall migrants, and did sunset point counts of clustering monarchs at Point Pelee. She was the eyes of the fall migration.

Point Pelee National Park is Canada's most southern tip of land. It's part forest, part marsh, part long spit of sand that seems to be redrawn daily by Lake Erie's moods. On my short visit I saw the lake's varied temper. From a serene, smooth-skinned surface to churning waves, the lake's frame of mind sculpted the park each day. As it transitioned, I saw my own reflection. I too settled with relief and riled with anger as optimism and pessimism kneaded my mind.

During my stop, Darlene and I visited the shores of Point Pelee several times. On each visit we sewed paths in the trees, our heads tilted to the shadows of the canopy. We thought perhaps we might see clustering monarchs. They gather each fall on the shores, waiting for favorable conditions, before crossing Lake Erie at its narrowest waist. Each autumn, Darlene tracks this concentration, an assembly preparing for the migration south. First they cross a lake and then they cross a continent.

Eyes to the sky, we knew it was too early. The silhouettes we glimpsed were only taunting leaves and tricky wind. It would be some weeks before Darlene would start documenting large numbers of migrants. She would keep me posted with daily updates. I wanted to see them, but I was also happy to have a head start. Mexico was still far away.

I had one more homestay in Ontario before beginning my own fall migration. I veered north, against a mild wind.

Louie Fiorino, a monarch enthusiast, traveler, and photographer, lived close enough to my last Ontario presentation that it seemed meant to be. As I turned onto his street, there was no doubt which house was his. A moat of plants, where a yard likely once stood, surrounded Louie's home. His driveway was an organized explosion of potted greenery. A small pond held hesitant frogs. Wary of winter's approach, they stood like statues until I got too close, then plopped into the safety of an icy bath. Like Louie, they were lured more by gardens than by grass.

After homemade biscotti, pizza, and, of course, ice cream, I rested soundly. The next morning, it was finally time to set my sights south. The monarch migration would soon be gearing up and I was feeling the pressure. My first step in heading south was crossing back into the United States, from Windsor, Ontario, to Detroit, Michigan. Easier said than done.

The bridge didn't allow bikes.

The tunnel didn't allow bikes.

The bus didn't allow bikes.

The ferry didn't allow bikes.

My options included a 126-mile detour north around Lake Saint Clair or a 140-mile detour south by way of an unreliable ferry. I had not made a plan because I assumed one would unfold as it always did. In this case, that presumption had been a mistake.

I once saw a drawing of pigeons riding bikes and the caption said, "Pigeons, like cyclists, must be content to live in the margins." The sentiment resonated with me. It was my life as a cyclist, explained. Being stuck on one side of the river with no options reiterated what I already knew:

society tells cyclists we're not important, that we don't deserve space. It wasn't just the lack of options that infuriated me, it was a system that demanded compromise from non-conformists. For me, bicycling was more than transportation. It was my version of praying. A way for me to give my energy to the world. An offering to my planet. Every day, I biked for the air, climate, frogs, and butterflies, and every day, I was told in so many words that my convictions were crumbs easily swept to the margins.

I felt the weight of society's indifference get heavier as my options to cross the bridge dried up. Then I got mad. I called the mayor of Detroit, asking for a ride and an explanation. I called the folks who operated the bridge. The mayor did not care, the bridge employees threatened to call the cops. My anger and sadness were seen as petty, my barriers insignificant. I was irrelevant, inconsequential.

Finally, I cried. I was mad that I cried, which made me cry more.

Meanwhile, Louie stood by, offering to drive me across the bridge. But I didn't want what should have been others' accountability to be erased by Louie's generosity. I didn't want the problem to be fixed with Band-Aids, so that the status quo could leave future cyclists in the same trap. I didn't want to be babysat and worried about. I wanted the Powers That Be to give me more than margins.

Scary mad, I eventually placed my bike in Louie's truck. I continued to boil with rage, resentment, and frustration as we crossed into the United States. A few days later, when I could think without drowning in fury, I would be ashamed of how upset I had been. Louie had seen me at my weakest, my most human, and I was relieved that he hadn't let me go through with my threat of a hunger strike.

The monarchs, of course, have been pushed much farther than I was, and yet they are determined to return, year after year. They do not keep score, do not fall into anger, do not lead with a grudge. They find people like Louie, Darlene, Bruce, Barb, and others across Southern Ontario and North America. These people quietly fill their lives with beauty and fight to give the monarchs space beyond the margins.

In Southern Ontario, I felt my kinship with the monarchs grow.

Headed
Back South

Five miles across the border, a bungee cord I hadn't stowed properly fell into my rear wheel. Hooked on a spoke, the bungee spun into my gears, where the cord and the gears fought, twisted, and pulled. Unaware, I was coasting toward an intersection, the drama below building.

Suddenly my wheel locked and my tire seized. Unable to spin, the wheel skidded underneath me. Crossing the intersection, my brain processed the moment as parts bent and snapped. My surprise became action. I jumped off, hoping to stop momentum and minimize damage. Too late.

The bungee cord had twisted my chain until the links had popped apart. The now-warped cassette (the gears on the wheel) no longer spun. The rear spokes, chewed nearly in half, threatened to snap. My lingering anger from the bridge crossing evaporated, as if the universe was saying, "Sara, you are so melodramatic. You want drama? I'll give you drama." I unloaded my panniers into a pile that looked as pitiful as my now-naked bike. Then, I laid out all my tools and began tinkering and testing.

Spread out on the sidewalk, my handful of tools looked like a mismatched football team: wrenches and Allen keys, a patch kit and pump, pliers, zip ties, miscellaneous screws, nuts, and bolts, extra chain, a chain-break tool, a chain cleaner brush, lubricant, and rags. Using the chain break tool, I removed the tortured chain and replaced it with a spare I carried. In denial mixed with hope, I tried to spin the wheel.

It didn't budge. The cassette was bent around the axle, and the wheel seemed to have admitted defeat. The combined problems couldn't be fixed on the sidewalk.

Because every problem is more of a problem when one is hungry, I ate my stash of leftover pizza. The black grease coating my hands added a light seasoning. Then I took a deep breath and ran through my options. I was lucky. It was a weekday, during business hours, in a city with a bike shop, and I had a phone. I called a bike shop to confirm their hours. Then I removed the chain so that I could spin the pedals without engaging the faulty cassette. I reloaded all my stuff onto my injured bike and took off down the road like a clown. I couldn't sit and pedal, nor did I have time to push and walk, so I stood with one foot on a pedal and used the other foot to push off the ground. It was like skateboarding, but with a bike. "No problem," I reassured my bike. I had a plan, and we were headed toward a solution.

I bikeskated through five miles of Detroit, got the needed repairs done, and ranted about my bridge crossing to a reporter who had called for an interview. Then I pedaled with purpose until both the bridge and the breakdown were far away. Once the day was finally done, I propped my rejuvenated bike against a lonely church, and laid down in the shadow it cast. Alone, I decompressed. After a tough day, did the monarchs, too, land in their nightly shelter and breathe a sigh of relief?

I awoke refreshed and headed out into my favorite landscape: corn country. It had been many miles since I had last been engulfed in the world of corn. The young starts that had just broken ground on my first pass had since grown into confident green stalks. The familiar feeling of desolation hit me. I had to admit, though, that it was nice to have a grid of farm roads. Traffic was light and my wanderings felt grand. At night, the corn was now a convenient curtain behind which I could shutter myself.

I camped in corn reluctantly. Between the nightmare scenarios of getting run over by monster farm equipment or getting poisoned by whatever kept all non-corn plants away, it seemed like a good idea to avoid it. Until I couldn't.

I pulled up to a school near sunset on my second night back in the United States. Aside from a smattering of small towns that grew like constellations in a universe of corn rows, the day had been spent in a monotonous mono-culture. It was easy to extrapolate and guess what was ahead.

The school was the first possible camping spot I'd seen in thirty miles. I inspected the quietest corners but found an exercise trail and construction project. A massive cottonwood sprouted from the surrounding cornfield. Its huge branches seemed like they could provide space underneath and a little protection. I jumped over a small ditch and slipped behind the first row of corn to investigate. Good enough, I figured, as I laid down on the ground, my helmet a pillow. Before setting up my tent, I took a moment to soak up the last of the sun and listen to the cottonwood leaves sing me and the corn a prairie's lullaby.

The endless carpet of corn parted to make way for Fort Wayne, Indiana. White-knuckled, in the growl of traffic, I found the University of Saint Fran-cis. I washed away three days of dirt in a bathroom sink. Clean enough, I changed from my only riding outfit into my only presentation outfit, then combed my hair, brushed my teeth, and called it good. I aimed for the line between clean and legitimate.

After introductions, I stepped onto the corpulent stage of the uni-versity's theater and told the story of my trip and the monarchs. Against a backdrop of dramatic religious canvases, I clicked through photos of snakes, toads, and caterpillars, the saints of my story. For an hour, the heaven above and the heaven at our feet shared the same pulpit. Through my stories, I attempted to acknowledge many heavens and to immortalize more than man. I wanted people to see the glory of swaying tallgrass, the grace of a slithering snake, and the forgiveness of a struggling caterpillar. I invited people to imagine that frogs, monarchs, and even tachinid flies are God.

In the audience, two nuns sat near the back and the monarch sisters, a group of friends devoted to monarchs, sat in front. From expert to novice, the crowd listened to my sermon. I felt supported and sure footed; neither lost nor alone.

Afterward I was ready to escape into a quiet edge along a rural road. But Louise, a professor at the college, invited me to stay at her house. I knew that between homestays I was using the anonymity of stealth camping as a crutch to ignore the weight of the world. Alone was a default habit for me—comfortable, easy, and safe. Alone, I was not lonely. To break the habit, I said yes.

Louise Weber, like all the folks I met at the Fort Wayne presentation, was full of ideas and action. She had written a new ecology textbook that helped students connect the dots between their environment and their lives. Though obvious, her idea was also revolutionary. Linking our actions to consequences means taking responsibility. After all, declining monarch populations are a symptom of a disease to which we are all exposed. Because they are indicator species, the monarchs and frogs and coral reefs are on the front lines today. It will be us tomorrow. Our only defense is to acknowledge our lifelines and to listen to the warnings of our fellow species.

The monarchs hold up more than the sky, they hold up the planet. They hold us up. Each creature is a part of Atlas, but our demands will topple the whole balancing act. This is not a threat, it is a fact. The way it is. Truth. My gut reaction is to crawl into a ball and close my eyes. Ignorance is bliss. Then I stretch out and realize the only way out is through action: expanding our collective consciousness, fighting with education, and forging connections to the people working hard.

It was thirty-seven miles as the crow flies, the butterfly flutters, and the butterbiker bikes, from Louise's house to Kylee Baumle's. A straight shot, over the invisible border between Indiana and Ohio. No customs, no agents, a line on a map.

It hadn't been my plan to bike those thirty-seven miles east to Kylee's. She had contacted me early on and had been instrumental in organizing my presentation in Fort Wayne, but her house was just a bit too far. Yet as my schedule solidified, I saw an opening to make the detour and learn from another key player in the fight for the monarchs. Kylee walked the talk—rearing, rescuing, teaching, protecting, and advocating for monarch butterflies. She literally wrote the book on monarchs, appropriately titled *The Monarch*. What was an extra thirty miles on a 10,000-mile tour?

Kylee was a monarch steward in every way. Her native garden was teeming with pollinators. Caterpillars rescued from a mowed roadside, now in Tupperware containers, ate their way toward migration. Honey made from milkweed nectar was spread on her toast. In her freezer she had the carcass of a parasitized caterpillar that she had encountered. The caterpillar had been killed by a tachinid fly, parasitoid flies that lay their eggs on monarch caterpillars. The eggs hatch, bore into the caterpillar, and develop inside, until they emerge to pupate. *Ouch!* I thought as I peered at the white maggots.

If Kylee's garden grew monarchs, then her passion grew awareness. After Kylee told the owner of a nearby field about the monarchs, he held off mowing, much to her surprise and delight. Her young neighbors would show up weekly with tales of their caterpillar-seeking adventures. Nile, the oldest of the crew, had the best review of my presentation: "I knew I was going to learn about the monarchs," he explained, "but I had no idea about all the comedy."

My arrival at Kylee's house was followed shortly thereafter by the appearance of two of my friends who happened to be driving across the country. Oh, believe me, I know, inviting friends to visit you when you are already a guest in someone else's home is bad manners, bad judgment, and all-around bad form—unless your host is Kylee. She pulled out enough food to feed the growing crowd and seemed in her element.

I hadn't seen my friend Matt Titre since we had said our goodbyes in New Orleans, at the end of a six-month river trip in 2016. He was traveling across the country with his girlfriend, Onthy Alexiades, whom I had not yet met. Seeing Matt was like returning to a well-loved home. He was a fellow adventurer and I didn't need to translate my stories. Telling them was enough for him to understand, without explanatory commas. We jumped into conversation as if our trip had never ended, and by the time we were all eating dinner, the absurdity of our reunion had grown into a comfortable, this-makes-complete-sense gathering.

Stories of our river trip surfaced, and I was struck by how different it is to tell stories in tandem than it is to tell stories by oneself—which I had been doing for the last five months. Relating as a team, feeding off shared memories, creating richness and drama together is impossible on a solo tour. Recounting adventures is a glue that creates comradery. I missed it.

Before this trip, I had never been on a solo tour for more than a few months. I wasn't sure I could handle ten months alone on the road, let alone pull off the logistics of presentations and media outreach. Scariest of all had been not knowing if I would like it. Trips are made by the people on them, and being alone meant it was up to me to make the trip. Every decision, every wrong turn, every scary road, and every strange encounter was mine. It was up to me, and without backup, I felt those pressures. But I also felt free.

Undertaking an adventure in a group is an intense endeavor. The highs and lows of travel demand that, as a team, honest conversations, realistic expectations, and scrutiny are always in ready supply. The group, as a whole, must welcome a similar brand of challenges and be ready to collectively meet the uninvited difficulties. In college, I found such kindred spirits. Together, we nurtured our collective craziness, and emboldened each other to reach beyond the status quo. I continue to count my wealth in the friends I have who are running toward the quirkiest of dreams, building models of possibility in the impossible.

Until my trip with the butterflies, almost all my adventures had been with teams like these. My role was always as the ideas person, the extremist, the boundary pusher. It worked only because I surrounded myself with patient people who could adjust to my varying degrees of intensity. If even one of them had been available, I would have turned my butterbike tour into a team mission.

But every year, it seemed that my pool of potential teammates shrank. As unconventional as we all had been, life was bringing order to most.

Marriage, kids, and normal jobs were taking the reins and setting the stage for new types of adventures. This solo adventure had been not an option, but my only choice. I felt like I had missed the memo, like an outsider among my own tribe.

A question emerged for me: was I rejecting societal norms because they were not for me or because I couldn't have them? The answer seemed to be both. Normal as defined by our culture seemed unbearable to me, and arriving there seemed as impossible as going to the moon. I had found only a handful of people that I could use as role models for what I wanted, and so I headed blind toward my unknown destination a bit bewildered. All I could do was shake off my doubt, look forward and continue on. I felt like I was at the edge of a cliff, and while going back would be less scary, jumping into the unknown and seeing where I landed was what I needed to do. Though it could feel like I was jumping alone, I was not. I was with millions of monarchs, the people taking care of them and me, and my friends cheering me on from afar.

Nearly six months in, I was enjoying traveling alone. But by the time I was at Kylee's in Ohio, I also knew that I missed having a team. Going solo, I never got a break from decision making, never had someone with whom I could commiserate or celebrate, and it was all on me to be a good guest.

Hosting me was an incredible kindness, which I tried to repay with gratitude, dish duty, and a few entertaining stories. This had been the norm for all my trips, but this time, alone, it was a harder task. If I was tired, there was no one to take up the slack. If I wasn't feeling social, there was no one to add energy and life to the stories. I couldn't pass the most common question along to my nonexistent traveling partners. There was no one to tell me I was being insufferable.

Shortly after dinner, I waved goodbye to Matt and Onthy. A few days later, I waved goodbye to Kylee. I looked up at the sky, hoping to see the monarch Kylee and I had tagged waving with each flap of her wings. Waving was the difference between biking alone and being alone.

With the Wind

MILES 6587-6804

"Holy sheep shit!" the man boomed. He leaned from his riding lawn mower to inspect my bike, while I patiently explained my trip and the contents of panniers. "This roadside could be *perfect* habitat for monarchs," I continued, the conversation started. My voice was light. He smiled a familiar smile. A smile that said, *Lady, you sure are strange.*

He wasn't wrong. I was biking thousands of miles with butterflies, picking up roadside spoons (a collection that is now in the hundreds) and roadside fabric (I like to make litterbug quilts). I didn't own a house, a car, or a pillow. I had a deep and seemingly unending reservoir of empathy for tadpoles. I *was* strange.

I waved goodbye; my work there done. He was likely to tell my strange story around the dinner table, and for a moment the monarch would have a seat, too. The more times he heard and told the story, the more likely change was to come. The rule of seven meant he only needed to hear about the monarchs six more times to remember, and perhaps start to change.

I held onto such hope, even as I moved forward alongside the ditches and their drama. Mowed ditches left me sorrowful, yet finding them spared was bittersweet. The caterpillars seemed so vulnerable, even as their presence proved them survivors. Their ancestors had navigated storms, predators, development, and disease. Each generation was a trophy of evolution. Yet it took just seconds for a mower to erase that which we look at but do not *see.*

All I could do was be strange and keep going. I had to trust that this mess all led somewhere. I had to trust that not all the monarchs I saw were fated to fall. I had to trust that hidden in the ditches, between the yellow petals and the fuzzy tufts of seeding clover, lived hope. Hope that confident fifth instar caterpillars would beat the odds. Hope that they would survive to pass me thousands of feet above, or sit quietly with me in Mexico.

If you don't mow, hope can bloom. Hope can fly to Mexico.

A blessing and a curse, seeing brought me grief, but it also brought me caterpillars. By Indiana, I could spot gorging caterpillars as I coasted by at twelve miles per hour. I could scan clumps of milkweed, their hearty leaves laced with pink veins, and note nibbled holes and eaten edges. Often the holes were lined with the milkweeds' white latex sap, the calling card of a caterpillar. I could even, from my bicycle, spot caterpillar frass.

Frass. I laughed when I learned that even caterpillar poo has its own name. Frass, scat, spraint, guano, cow pie, horse apple, manure, dung, droppings, castings—scientists have more names for crap than a ten-year-old boy. Call it what you want, by the time caterpillars reached the third instar stage, the evidence of their ceaseless appetites—small, green nuggets—was often more detectable than the caterpillar itself.

A few days south of Kylee's house, I stopped to investigate a milkweed. Among the leaves sprinkled with frass was a partially eaten leaf hosting a large, confident caterpillar. The brazen fifth instar ate down the edge of the leaf like an inspired writer pounding a typewriter. Then, surprising me, the eating machine paused to clean up a mess of its own making. The plump caterpillar turned and bit down on a nearby drop of fresh frass. For several moments it held the frass in its mandibles, swaying its upper body like a crane out of control. Then it flicked the frass over the edge of its leaf, where the waste joined a growing collection, before the insect resumed its eating chores.

Another layer revealed by nature. The discovery of an intriguing detail drowned the dread of the news, the facts, the future. My worries, for a moment, could float away with the wind.

The wind was my constant traveling companion. Whatever my direction, the wind had its own. Sometimes we matched, sometimes we collided. No matter which way I pointed, the wind set the tone. The wind was king, the decider, the difference between merciful and punishing days. Leaving Centerville, Indiana, the day after a presentation, I looked at the forecast for a prediction of the wind's wrath. The weather report seemed to think that a coming storm would bring one day of tailwinds, then pivot, for three days of headwinds. A plan formed: I would exploit advantageous winds by biking as many miles as I could that first day, traveling with the storm, and let it push me south toward my next stop in Evansville, Indiana.

I woke up early, before the rain clouds cried, and was pedaling as the storm winds brewed. The hills unrolled, and by lunch the rain had caught up. A moving puddle, I was warm as long as I did not stop. The tailwind was my coach, patting me on the back and keeping me focused. Go, go, go! it seemed to say. Mile after mile, I let the wind and my legs work together. By the time the sun set, the storm had cried itself out, but I was ready for more. The setting sun ushered in the magic of darkness and the creatures at home in it. While all the monarchs rested, I could gain ground. Sure, the monarchs could soar thousands of feet in the air and take advantage of air currents, but I could travel at night.

The Indiana darkness was warm and humid—fading remnants of the storm's visit. Without ceremony, I switched on my lights and accepted darkness's invitation to join the nocturnal world. The powerful white beam of my front light swam through the moisture rising off the pavement, and I found myself hurtling (slow feels fast at night) through the mysterious fog.

I could feel the ride: temperature changes at creek crossings, the quiet whir of wheels embracing road, the measured breath of ups and downs. My brain tried to translate what I could not see. The darkness beyond my light ensured that only the blink of now and the small sliver of just ahead were relevant. For a few miles I had no past nor future. I was simply biking with the night.

It was the shadow of a salamander wandering unsuspectingly into the land of the paved that broke my trance. I slammed on my brakes, tucked my bike indelicately into the ditch, and ran back to investigate. At night, quiet and darkness are like guards. I could detect distant cars by cluing in to the glow of their lights and the rumble of their engines long before they were threats. I could give my complete attention to the salamander's marbled skin, its frantic gait. Squatting low in the remoteness of night, I looked into its eyes and saw my own.

I moved my fellow creature off the road.

My eyes were trained to register the shadows of hidden, camouflaged, and reluctant amphibians. I could tell leaves and springs and broken tires from hunched frogs and oblivious toads. It was what made me a scientist, but it was also what broke my heart. When I found them already dead, I stopped to move their lifeless bodies off the road, and apologized by providing a more dignified resting place. When I found them dying, I ended their misery. Toads leaking guts cried blood. I cried tears.

Perhaps I am crazy to feel such anguish for a toad, or perhaps those of us who do are the last of the sane. Either way, I let each death touch me, absorbing the stillness so that the deaths don't go unnoticed. One day, we will look back in shock at our relationship with other living things.

I moved the salamander, still alive, off the road, and hoped that it would live out the rest of its life away from that road. I traversed Indiana under a sky of stars, and felt my luck with each amphibian sighting. I relocated a grey tree frog with skin smooth to the touch, but rough like bark to my eyes. I scooted a toad off the road. In the shoulder, he continued to stand sentinel, stout and full of attitude, with eyes that glittered gold. Another marbled salamander, black and white, crawled close to the ground and reflected the Milky Way. I chaperoned her walk to the wet leaves of the ditch.

Though I wanted to bike all night and connect with the new day, my legs grew too tired, my wet feet started to complain, and my eyes longed to shut. A few miles over one hundred, I said good night to the frogs and the darkness. I had gained my ground and taken advantage of the tailwinds. Now it was time to rest. If the next day brought a headwind, as the weather report predicted, I would need my strength.

As soon as I decided it was time to camp, a spot appeared. A church with a water spigot, working outlets, a sheltered pavilion, and a covered, out-of-sight, tent-sized cranny at the back of the building welcomed me. It was perfect. A bike tour miracle. Just like tailwinds, the simple victories of finding water and shelter were powerful. Warm and dry, the rain bouncing lyrics on the roof, I appreciated the win.

Maintaining
the Lead

Six months before I arrived in Evansville, Indiana, Val Alsop and I had met at the El Rosario monarch sanctuary in Mexico. After exchanging contact information, Val had spread the word to her community in Evansville, and Gena Garrett had contacted me about giving presentations there. That was enough to steer my route, and 6945 miles from El Rosario I rode into Evansville. The sun reflected off the pavement as I reflected on the magic of connection. Like the monarchs, Val connected Mexico to Indiana.

Gena had scheduled a full lineup of presentations and a place for me to stay. I was the sole visitor at the University of Evansville's guest house, free to listen to my music loud and work out logistics till the wee hours. The fridge, a novelty, let me splurge on foods that bike touring never allows: cantaloupe (because it is heavy and too much to eat in one sitting) and ice cream (because it melts). Simply using a cutting board, rather than my cooking pot lid, was luxurious.

I made presentations at several schools and the university, as well as to a youth group called the Navigators. All the Navigators brought their bikes to my presentation and afterward, the boys and girls of varying ages sorted out their steeds from a chaotic pile and donned their helmets. Together, we set off for a local pollinator garden. We were a critical mass of junior butterbikers.

We were also a critical mass of chaos. At first, the tiny bikers careened nearly out of control down the bike path. All I could do was pick up the kids who fell (nothing serious) and twist loose handlebars back into acceptable positions. Eventually everyone found their balance. The adventurer inside each of them grew one ride stronger.

On our adventure, I saw myself in several of the older girls who smiled as we hooted at the setting sun. My youthful craving for freedom, challenge, and adventure was theirs. When I was young, I would ride my bike until I was lost. Places I knew only by car and with adults would reveal themselves to me. The realization that I could get where I was going on my own, under my own power, unlocked the bigger world to me. I knew my town in a way the other kids didn't, and I carried that tiny glimpse of freedom with my head held high.

I cheered with the young girls. They were finding what I had found, what I still sought each time I got on my bike: to be part of a pack, a flock, a pride, or to be a lone wolf running toward the horizon.

I left Evansville with Susan Fowler, a woman with enough energy to move a mountain. After meeting her at one of my presentations, I quickly learned that she had a heart of gold (she showed me her impressively large "ball of smiles," made from strands of yarn) and an unending love of conversation. The morning I was leaving, as I worked quietly over my computer, I looked up to see Susan standing outside, talking to me through the window. She wanted to know if I had received her 5 a.m. text message explaining the season's constellations and describing every grocery store I might encounter on my way out of town. Wide-eyed, I reminded myself that her energy was better than apathy. Her voice was bringing the world to a better place. I nodded yes.

Susan biked with me out of town, to a bridge under construction that loomed intimidatingly between Kentucky and me. I had been warned about it, but swimming across the Ohio River was not a real alternative. Susan gave me a half empty container of Eclipse gum and waved me goodbye.

As I approached the bridge, I took a deep breath to prepare mentally for the quarter mile ahead. Construction had turned the bridge into a funnel.

One lane was closed, blocked by cement barricades that constricted what was left. A car could barely fit, let alone a car passing a bike. The only way to deal with the bridge was to use my body as a barricade so nothing could pass me going up and over the river.

I waited for a break in the herd of cars and pedaled with every ounce of strength I could muster in an attempt to keep up with the traffic going relatively slowly through the narrow passageway. Looking over my shoulder I saw a huge semi-truck approaching behind me, closing the gap. Rationally, I knew that the semi could see me and would not run me over, but I still looked back with fear. The gap grew smaller. I pedaled faster. I cursed every road authority with the audacity to leave safety up to me. I cursed the constructions workers who yelled, "Hey that's illegal!" but gave no clues as to what I was to do. Swim? Detour 200 miles? Move to Evansville? There were laws to protect the men that yelled but only luck to protect me (besides, it was not actually illegal).

I raced ahead. The semi grew bigger, but I had nowhere to go. I dug in and didn't stop until the bridge ended and a real shoulder appeared. I veered toward it with panicky gratitude. My muscles were free to rest, my heartbeat could slow. Miles later, my lungs still burned, and the next day my legs were sore.

But I was alive and in Kentucky.

My first stop in Kentucky was at the edge of a field bejeweled with the orange blossoms of butterfly weed (*Asclepias tuberosa*). This milkweed is especially hairy. Monarchs actually prefer laying their eggs on the flowers instead of the leaves of butterfly weed, perhaps because of all the fuzz. *Asclepias tuberosa* is the poster child of the milkweed movement. Its clusters of shooting stars are like small fires burning sporadically in the green of its waxy leaves and shier prairie neighbors. Unlike a few milkweed species, which spread both by seeds and by rhizomes, butterfly weed depends entirely upon its pods of seeds and the wind to gain new ground.

Such pods, and their explosions of seeds, have delighted kids and the wind for generations. They appear each fall from the milkweeds' umbelled flowers. As the last green of fall fades away, the teardrop-shaped seedpods eventually dry, crack open, and expose a fluff of feathered seeds. Each milkweed species has its own style of pod; the butterfly weed's is long and bulbous, its drying husks are smooth and brown. The seeds, crowded inside, wait for the wind to tickle their flossy parachutes. Like miniature skydivers, the seeds are picked up by the wind and carried by their floss (also called coma). Thanks to variations in seed mass and coma lengths, seeds can be carried a wide range of distances. This increases their distribution and the opportunity for at least a few seeds to settle in open patches of ground, where they will find the most success. All will settle, but only those lucky enough to land in select spots will grow.

Though the wind is responsible for dispersing milkweed seeds, people have begun lending a hand. When the pods begin to split open, the seeds are collected, separated from their comas, and dispersed strategically. Some seeds are scattered in backyards, others germinate in pots destined for parks and schools, and some are rolled into balls called seed "bombs." Does the wind get jealous when seeds are combined with clay and compost and sculpted into seed bombs? *Nah,* I figure. Even the wind sometimes needs help.

Regardless of the courier, the key to germination for native milkweeds and other temperate plants growing in the monarchs' summer range is for the seeds to be subjected to winter. After experiencing both cold and moist conditions, known as stratification, the seeds are spurred to germinate. Some monarch enthusiasts trick their collected seeds by storing them for the "winter" between moist paper towels in the refrigerator. Other gardeners plant seeds in prepared beds or containers in the fall, and let real winter do its thing. In the spring, after germination, the sprouted seeds can be transplanted.

The requirements for milkweed to thrive vary greatly, even among the same species. This is because of the variety of conditions to which they have adapted. A seed from a common milkweed local to Kansas will be more successful when planted in Kansas than if planted in Ontario. And vice versa. Local plants evolve localized adaptations. By sourcing seeds from plants in your area, the adapted advantages they harbor will tip the scales in their (and your) favor.

A few milkweed species spread by rhizomes. Rhizomes are essentially underground stems. They allow plants to send runners out and colonize new ground. Common milkweed (*Asclepias syriaca*) is the king of rhizomes, which makes it an adventurous explorer in my eyes, but an unwelcome guest to others.

The butterfly weed growing along the road did not employ rhizomes, and did not yet harbor seedpods. The flowers still shone bright, seeking pollination from the rainbow of life they hosted. Two fritillary butterflies, with their dull cloaks of checkered orange, sipped and swigged the butterfly weed's nectar. Like brushes being dipped in paint, they seemed to guzzle the flower's shine with their proboscises. A caterpillar I didn't recognize lay camouflaged on a twig, visible only by the honey-orange dots of its crown. A faded and tattered swallowtail, missing at least 40 percent of her wings, landed stoically nearby, all of her yellow-orange scales stolen by time. I saw a splendor not found in young, vibrant butterflies. I saw beauty in her saga and hope in her clinging color.

A male monarch swooped onto a flower, the orange of his wings and the orange of the petals reflecting each other. Orange was never just orange. Mother Nature employed many hues. Like a paint store sample card, "Tangerine Carrot" mixed with "Auburn Marigold," and "Ginger Flame" danced with "Ochre Sunset." *Danaus plexippus* orange and *Asclepias tuberosa* orange were rendered by Earth's algorithms.

My schedule and that of the monarchs pulled me out of my trance. Looming presentations and winter's persistent chase were catching up to me, to us. I felt the warnings in the cooling nights, shorter days, and browning countryside. My phone buzzed. It was an email from someone in the north letting me know that the monarchs were starting to migrate. I needed to hurry. I didn't want them to pass me before I hit Missouri and then have people telling me for thousands of miles that I had just missed them.

I can't say if the monarchs felt the same urgency I did. Perhaps for them it was a mere craving, a hankering, a pull. For me it was alarm bells and sirens. I felt like I was slowing as the miles grew. Mexico seemed to be getting farther away.

I stopped to watch a small female monarch. She hovered by a stand of common milkweed. Both looked weary in the heat. I watched as she landed on a leaf, curved her abdomen underneath, and laid an egg. I was amazed. I had thought that if tagging was starting, if milkweed was dying back, and if fall was approaching, then the migration had begun. That egg hinted at something different. It told me I had time.

All of a sudden, I began to find eggs again. Each egg, needing around a month to develop into an adult, meant I had a month to get a head start (though as I learned, it is possible that those eggs were simply born late, and didn't have time to make it to Mexico). To understand why that female was laying eggs this far south, imagine it is early August and you have just eclosed from your chrysalis in the far north of your range. It is too early for you to be a migrant, so you will mate and lay eggs. Your eggs will metamorphose in thirty to thirty-five days, and your kids will fly to Mexico (congrats). If you lay your eggs at the same latitude at which you were born, then your kids will become migrating adults in mid-September, a time of seasonal and potentially deadly freezes in the north. So instead of laying your eggs in the most northern parts of the range and risking a killing frost, you move south, on a mini migration, to lay your eggs. As previously mentioned, this is the season's third, or midsummer, migration (also referred to as the pre-migration).

Those eggs had a bit of a head start, away from the cold and closer to Mexico, but not all monarchs did this. Flying south was a gamble. You might avoid the northern cold, but in the south, you would find waning milkweed. Since each season was different, the population hedged its bets, with some staying and some pre-migrating. Only time would reveal which option was best each year.

The race was on, but it was not yet a sprint.

The
Out-of-the-Way
Way

I crossed into Missouri at Cape Girardeau on a bridge that spanned the Mississippi River. New by bridge standards, it was wide with a generous shoulder, and I could stop at its crest to peer into the churn. Below was a river of rivers. The Missouri, Yellowstone, Kaw, Illinois, and thousands more waterways, all swallowed into one. I could hear the stories of every tributary mixing with prairie mud, mountain rock, human folly, fish, foam, trash, and gravity as the flow proceeded toward its ocean dreams. The Mississippi yawned wide and the bridge had to leap to reach from shore to shore.

Cars rushed by, and I contemplated the unique collection of storms, winters, and springs that gathered and flowed here—just as the stories of each season's monarchs came together in their migrations. Every day saw a new river, just as every year saw a new migration.

I left the river and rose with the land out of the flood plain to Cape Girardeau, where there were presentations and two nights of hospitality.

Leaving Cape, as the locals called the small city, I was on the lookout for a Burger King. I had been given coupons for free soft-serve ice cream after a presentation, and I was looking forward to cashing them in. I was also on the lookout, as always, for milkweed. I scanned a field alongside the road. An industrial mower was slowly and deliberately cutting all vegetation—and unknowingly destroying monarch habitat. I dropped my bike and ran into the field, methodically searching for caterpillars on the doomed milkweed. I worked quickly. The mower crept closer; the cemetery of loss expanded.

On a small milkweed, I found an unsuspecting caterpillar. While it couldn't sense the approaching mower, it sensed me, and stopped eating to search for answers with its waving tentacles. I yanked the milkweed and its caterpillar tenant out of the ground and tucked them into my pannier. Then I headed down the road, on the lookout for the perfect relocation spot.

I don't know what the caterpillar was thinking as it bobbed in my pannier, but I know what I was thinking: how disheartening it was to help-lessly watch the mower decimate habitat and destroy possibilities. I was thinking about my new charge, whose tentacles now stood like snorkels at the edge of my bag. Heavy on the brakes, I tried to make the ride less terri-fying for the refugee. I avoided bumps as I evaluated the roadside habitat. Not just any milkweed would do. Most were either too close to the road and risked being mowed, or too old and yellow. Finally, six miles down the road, I found the perfect spot. It was a ring of milkweed surrounding a util-ity pole. I could see a resident caterpillar, which ruled out excessive toxins. Gently, I placed my caterpillar transplant on the leaf of a vacant milkweed. Wishing it luck, I carried on.

Honestly, most of my monarch sightings run together, but that cater-pillar will always stand out. I wonder now, did it fly to Mexico and add its story to the archive of butterflies that somehow beat the odds? Did a differ-ent mower cut that dream short?

Caterpillarless, I turned onto a dirt road that squirmed indirectly across the map. It was not the most practical route, but the demanding climbs and white-knuckle descents helped motivate the undertaking. Lost in the moment, I crashed through a creek crossing and emerged on the other side with waterlogged shoes. A section of washboard road rattled my brain. It was nice to be so consumed by the terrain that I couldn't think about how many miles were ahead of or behind me. That night, I waded into a small creek alongside tadpoles and reflective-eyed spiders. The coolness of an Ozark creek and the obsidian darkness of the sky swallowed me wholly. Or was it holy?

The backwoods nights were part of the answer to a question that was swimming somewhere in the back of my head. *Why*, it quietly asked, *am I going north?* North was the opposite direction of where I was headed. Like an oxbow in a river, my route was not straightforward and seemingly foolish. But I had a plan. Though my current route added miles and at the moment was taking me in the wrong direction, it would also allow me to jump on the Katy Trail, the longest rails-to-trails bike path in the country. Though my question was real, it was overridden by my desire to get to the Katy Trail. For me, interesting detours could actually feel like shortcuts.

I rode the Katy Trail and its spurs some 250 miles toward Kansas City. Gentle grades and living windbreaks complemented the trail's crushed limestone surface. Dappled by sun and shadows, the trail also attracted a shocking number of basking snakes. The potent heat of a fluke weather system seemed to have lured out even the wariest of creatures. Some snakes slithered off before I had a chance to give them my full attention. Others stayed longer and became teachers.

The snakes I caught and photographed linger longest in my memory. I can still feel the quilt of a rough green snake's keeled scales twisting through my fingers. I see her eyes, huge and alive, watching me. A tube of muscles, a balance of power and grace. The peaceful beast, a beautiful Jolly Rancher–green, melted into the leaves when I let her go.

I let the snakes that I caught go, but took pictures of them with me. The encounters were no doubt terrifying for them, but their images allowed people to look into reptile eyes and evaluate their own fears. It is easy to be scared of something we have never seen but have learned to be afraid of (green tree snakes are harmless). Snakes, bears, bees, wolves—we are taught to fear wildness. Wild spaces also cause anxiety. We let fear control us, leading to an unofficial but no less real war on nature, on the untamed.

The trail, with its shady tree tunnels, offered temporary protection from an *actual* existential danger. Across the Midwest that year, late-September heat waves were shattering century-old records. In the United States, 2017 was on track to be the third warmest year on record (2012 and 2016 were warmer). The monarchs were coping as best they could, but climate change was heavy on my mind.

Even if the monarchs and milkweed plants are able to shift their ranges northeast as weather becomes more extreme, variables in the delicate migration equation—from nectar quality to wind patterns—could easily become too disruptive and lead to the migration's collapse. Moreover, the overwintering forest in Mexico can't "adjust" northward. This critical habitat is projected to disappear by 2090.

Knowing such realities, we can't excuse inaction. We must rise up, with the oceans, and demand a future. Children and young adults are protesting, technology is advancing, gardens are being grown. In Mexico, scientists such as Cuauhtémoc Sáenz-Romero are giving the monarch overwintering forest, and the monarchs, a helping hand. Because oyamel firs growing even 1000 feet apart in elevation are genetically distinct, Sáenz-Romero is collecting lower-elevation oyamel seeds (more tolerant of heat), and translocating them as seedlings 1000 feet uphill (to areas that will soon need more heat-tolerant trees) for reforestation, helping them compensate for the warming climate. By taking such measures, he is preparing homes for what will hopefully be future generations of monarchs.

With a deep breath, I pushed back the dread of climate change and focused on celebrating the creatures that are still here to save. Besides the basking snakes there were box turtles, like tanks, charging across the open path. At my side, congregations of sunflowers prayed to the sun and pollinators graced flowers. In the midst of the festivities I noted a monarch flying directly at my side. For just a moment, our cadences matched and we were traveling toward the same horizon, like a team. When the monarch changed direction with the wind, I stopped to record the time and location in my notebook.

At first I assumed the next one I saw was the same butterfly, returning to cheer me on. Then I noticed a second and third flash of orange. I kept my speed constant and watched a fifth, a twentieth, a fiftieth, and a seventieth monarch intersect my path within minutes. Some swooped in pairs above me, as if dangling a treat just out of my reach. Others passed perpendicular to me, wasting no time getting to where they were going. I counted and watched and admired the performance.

I took this increase in monarchs on the wing as a sign that the true fall migration was gearing up. Unlike in spring and summer, the fall migrants tend to cluster each night in roosts and disperse each morning, as if training for their winter repose. There are other signs when fall migration is near, such as butterflies nectaring intensively to bulk up, and groups flying in a uniform, southern direction. I began looking to the skies and the flowers for more hints of fall.

Beyond butterfly migrants, the Katy Trail connected towns, cities, and people whose work inspired my miles. En route to Jefferson City, Missouri, I stopped to visit Joe Wilson on the banks of the Missouri River.

Joe, a retired boxing coach, dog protector, and defender of the Missouri River, had once told me that he had only three rules: you don't hurt women, dogs, or children. If you did, he would deal with you. His stories portrayed a man with a steady fist and gentle heart. He spoke his mind and never wavered in his convictions. I saw my own intensity reflected in him, and it took only a few conversations to know that we were kindred spirits. We would both "buy a bucket with a hole in it for the right price." We would both support the creatures that needed us, regardless of what others thought. He was one of the Missouri River's great advocates.

That river—my river, his river, the Missouri River—had wrangled our introduction. Headed to the gulf via canoe, friends and I had camped several nights with Joe on the sandy stretch of beach he had adopted and protected with a fierce loyalty. He did not own the property, but that small piece of public land seemed to own him. They had adopted each other, and

I never asked how it had come to be. Sitting around a fire with the sound of the river, passing trains, overhead bridge traffic, burning wood, and Joe's voice may be my most sacred memory of that entire river trip.

Another memory: after deciding to bike with the monarchs, I visited Joe and told him about my idea. While he listened, a monarch drifted by. It was the first monarch I had seen since committing to my plan, and it felt like a blessing.

I missed my friend, the cheerleader, the fighter. Joe had died before my trip started. His stretch of the river felt empty without him. I missed his passion, his rough edges, and his genuine love. His authenticity. If all of us committed to one footprint of land like Joe had, the world would be a better place. I picked up some trash, leaving my silent tears to sink into the sandy shore and drift with Joe's memory in the river's muddy arms.

Upstream, in a residential neighborhood of Columbia, Missouri, I visited another person committed to a small plot of land. Around Nadia Navarrete-Tindall's house, a garden of native prairie grew. The only reminders of suburbia were the sidewalk and driveway cutting through the fortress of plants, and the neighbors' green lawns. I was thrilled to arrive at such a collage of color, texture, and life.

Just a few feet from Nadia's mailbox, ten mating pairs of walking sticks swung like trapeze artists from their actual stick homes. Below, a monarch caterpillar gorged. In the space between, bees, wasps, and skippers delighted in the many meal offerings. Each and every animal I saw existed entirely because Nadia gave them space to live. Those animals owed their lives to her generosity, her commitment to nature, her stewardship, and her bravery for ignoring the pressure to have a typical green lawn.

She invited us all to see a new way. She invited us to see our yards as something more. Beyond the property line, a few clumps of common milkweed stood in the neighbor's turf, out of place, like brave scouts forging ahead to scan the horizon for hospitality. Nadia explained that her neighbor no longer mowed down the trespassing milkweed after learning that it was the monarch's only chance. Knowledge, like the rhizomes of persistent milkweed, spreads slowly but surely.

That triumphant milkweed stood alone, doing its part. But it was not alone. Many of us were trying to spur action. My mission was to be just such a catalyst of change.

While I was visiting Nadia, I made a side trip through a serious rainstorm to the University of Missouri's A. L. Gustin Golf Course. The fairways were typical: high-maintenance, non-native grass. But unlike other golf courses, bordering the fairways were roughs composed of prairie and woodland rather than the traditional tall-cut grass. Even in the rain, the roughs popped with color.

Isaac Breuer, the grounds manager and my undeterred-by-the-rain tour guide had used nature to help create his vision. First he had removed the invasive honeysuckle. This allowed the waiting, native wildflowers to return. Slowly, acres of golf course roughs reverted to native prairie, which saved Isaac about $300 per acre each year in maintenance. Work for the earth, and the earth will return the favor.

The unconventional golf course was nurturing not only native plants and pollinators, but stewardship in unlikely circles. Golfers came and saw a different kind of example. Many at first complained about the "weedy" look, but Isaac was patient. He took the time to talk to each confused golfer, helping them understand and appreciate the native plants. Isaac was a voice for the monarchs, reaching people that I likely couldn't.

Nadia talking to her neighbor; Isaac talking to golfers—every conversation adds up. No one can talk to everyone, but everyone can talk to someone.

Ten miles from Kansas City, my bike surrendered.

The rear tire exploded and suddenly my rim was plowing gravel. I stopped and found a six-inch gash in my tire, not a normal puncture like I had hoped. The gash had likely been there for some time, getting weaker and weaker until the interior tube, full of air, had squeezed out and under much pressure, had popped. I took off the tire to repair the gash with a piece of rubber on the inside wall. Then I put in a new tube, filled it back up with air, and returned it to my bike. Problem solved.

For about two minutes.

The gash was too big. The tube again began to squeeze free, the rubber patch barely holding. This time I didn't wait for another pop. With no bike shops nearby, it was time for tire triage. Rear brakes disconnected, I sorted through my collection of zip ties (nearly all gathered from the side of the road), and fastened a handful of them around the tire, pulling the gash together like stitches. Operation Zip Tie left me without rear brakes and with a bump I could feel with each rotation of the wheel. Luckily, it also left the bike with just enough life to limp slowly through the suburbs to my parents' house.

I arrived ready for a break. My family knew I was coming. My mom had prepared her spaghetti-squash-casserole specialty, and my dad had organized a small family gathering. I granted myself one day to not think about biking, monarchs, miles, or schedules (at least not too much). One day to sit on a couch and eat. As satisfying as it is to get plenty of exercise and cross items off to-do lists, doing nothing from time to time is necessary and satisfying, too.

The next day, still taking a break from biking, I got to work on the hidden parts of my trip: the interviews, updates, and planning. From Kansas to Arkansas, I had already worked out my presentation schedule. That just left Texas, specifically Austin, where ten schools were interested in hosting me. I pored over maps, scoped out a potential route, gave possible dates to a dozen contacts, and sent out feelers for places to stay. Austin was 800 miles away. As long as I was headed south with the monarchs, I had faith the details would work themselves out.

Straggling
with Stragglers

By late August, the midsummer migration had bled into the actual fall migration. The metamorphosing caterpillars—taking cues from the shortening days, cooling temperatures, and aging milkweed—were eclosing in sexual diapause. The monarchs were assembling into a directional stream and flying with purpose. The overwintering generation was heading south.

Each year, the fall migration begins in the monarchs' northern range, a rally of monarchs moving south. This main mass has a leading edge, which is followed about a month later by a trailing edge. As the migrants advance south, they encounter recently eclosed monarchs geared up to go. These new recruits join, and the wave of monarchs grows as it moves south. The pace of the migration's progress seems to be a function of the monarchs' daily velocity, seasonal temperatures, and the angle of the sun.

Each day, the sun appears to be thrown across the sky like a fiery ball. At sunset and sunrise, the sun is at its lowest point, or zero degrees. The daily high point, at the apex of its trajectory, is called solar noon. This angle changes daily. In Kansas City, for example, the sun at solar noon ranges from 74 degrees on the summer solstice (its highest point) to 27.5 degrees on the winter solstice (its lowest point). The fall migration appears to track some variable related to this changing angle of the sun.

The leading edge of the migration begins traveling south when the sun's angle at solar noon is 57 degrees. Higher latitudes see the sun at this angle sooner than lower latitudes. In Minneapolis, this occurs around August 21, but going south, the date is later and later: September 6 in Kansas City, and September 29 in Austin. Amazingly, the first monarchs arrive at the Mexican colonies at the end of October, when the sun at solar noon is at—you guessed it—57 degrees. Furthermore, the trailing edge of the migration corresponds latitudinally with solar noon reaching 46 degrees. In Minneapolis this falls around September 19, in Kansas City around October 4, and in Austin around October 29. Wild-caught monarchs that were tagged in this window, when the sun's angle at solar noon was between 57 and 46 degrees, have accounted for 90 percent of the monarchs recovered in Mexico. This suggests that monarchs migrating in this window have higher survival rates.

Some years see a higher percentage of monarchs arriving outside this solar noon window, though always later, never earlier. From 1994 to 2018, four of the five lowest overwintering populations were associated with a high percentage of late arrivals. There are certainly many factors at play, but in general, pacing with the sun seems to help monarchs successfully arrive in Mexico. Though the exact cues remain unknown, the sun's angle is part of the soundtrack to which the monarchs dance. If their pace can match the rhythm of the sun, then daylight, temperature, and myriad other factors seems to be in their favor.

When I left Kansas City on October 11, I was with the stragglers; the sun's angle at solar noon was 43.6 degrees. Still, the monarchs I saw held much promise as they piled onto bowed flowers and took long pulls to satisfy their nectar cravings. Guzzling nectar is a key part of the fall migration. Each monarch needs to not only navigate thousands of miles to a forest they have never been to before, but they need to arrive as fat as possible. They need to build up enough lipid reserves to survive the winter in Mexico.

Though the monarchs' winter home in Mexico is graced with bouquets of white, purple, and yellow flowers, there is nowhere near enough nectar to power every migrant. The majority of monarchs rely, instead, on fat reserves to survive. Only when their fat reserves become depleted do hungry monarchs coat the blooming scaffolding of the understory. These nectaring monarchs have been found to have half the lipid reserves of their inactive counterparts in the hanging clusters. These are the monarchs that didn't stockpile enough fat during their fall migration south, perhaps because they were in a rush to arrive, had found poor quality flowers, were born small and thus had a less efficient glide ratio, or any of a slew of other reasons. Whatever their situation, they are desperate for a midwinter snack.

Unlike nearly every other migrant on the planet, monarchs need to arrive at their destination heavier than when they start. I would likely be arriving heavier, too. If I hadn't encountered such wonderful (and delicious) hospitality, or encountered so many grocery stores, or biked faster than my slow and steady pace, I am sure I would have lost weight. As it was, my steady (and surprising) gain was another thing I had in common with the fall-migrating monarchs.

Leaving Kansas City, I knew I had more to do than eat my way to Mexico. I had eight days to pedal 280 miles and make presentations in six cities along the way.

The time crunch meant that I arrived at Val Frankoski's house in Joplin, Missouri, in the dark. It wasn't until the next morning that I could see Val's native garden, which stretched across her backyard and dominated the border of the front yard. When I did, my visit coincided with one from a monarch, confirming Val's dedication. Within two days, thanks to Val's passion and organizational skills, I was able to make presentations to over 1000 students. I also got to visit some of the community gardens Val had organized and promoted, which had hosted monarch caterpillars all summer. If only one caterpillar survived, if only one child was inspired, that was enough.

Val saw me hopeful, telling kids about the joy of monarchs, adventure, and science, and how everyone can be part of the solution. She also witnessed a low point of my trip, when I hit my breaking point, and despair seemed to gain the upper hand.

It started as a friendly conversation with an older man. Val saw we were on a collision course and tried to intervene. Unfortunately, our discussion was fueled by politics and different perceptions of reality. He thought the only difference between success and failure was a willingness to work hard, and that it had nothing to do with a system whose rules were written by people that looked like him. I believe those rules cleared his path, while leaving others trapped by dead-ends.

I tried to point out our shared advantages. I wanted to stay calm, listen, and reflect, without grief controlling me. People told me I was strong because I could bike for months, organize, and fight, but in many ways, I felt weak. I struggled to find compassion. Most miles, most days, I struggled to forgive those destroying the prairie, erasing great migrations, and acknowledging neither.

Our dispute derailed. He could not, nor did he want to see, that simply being a white male was enough to make his life easier. Financially, socially, and educationally, the deck was stacked in his favor. I emphasized that his successes were not to be diminished or lessened. But I also pointed out that in recognizing his own good fortune, he could also acknowledge the disadvantages of being dealt a more challenging hand. His closing arguments were well worn and unconvincing: success was simply a product of hard work.

The monarchs knew this was a lie. They worked hard—flying thousands of miles, braving storms, predators, cars, mowers, and disease—and still their population crashed.

I also knew this was a lie. By the time I met that fellow, I had been working so hard that my body no longer felt like my own. It was merely a tool to get me from one presentation to another; to wake people up; to absorb the destruction of the world. I was working *very* hard, the hardest I could, but the monarchs' future still seemed so precarious. That man had handed down a broken world and acknowledged no responsibility.

He was my last straw.

Val did her best to reassure me as I walked away, too tired to talk any more. I was spent, overwhelmed by the apathy of my country, the lack of compassion I saw each day, and the vileness of powerful people stepping on the weak. I told myself to be strong. I too could ignore the problems not yet touching me, but I didn't want to be that person. I knew we all could, and must, do better. I knew seeing ourselves as part of the problem was the first step to being part of the solution. I was far from perfect, but I was trying.

I kept presenting. I kept biking. I left Joplin just as I had arrived, in the cover of night.

In the dark, I pedaled to Neosho, Missouri. It was a misty night, which brought out the frogs and my amphibian spirit. A small leopard frog hopped toward me, as if asking to be scooped off the road and thrown to safety. I stopped but was a few seconds too late. A car sped by, giving me no room, and another innocent frog disappeared. Erased by the car's tire. Forever. From a living creature to a spray of body and rain, without even a moment of recognition. No hesitation. No sorrow. No frog. A strange, calm silence on the highway.

Most animals cannot fight back with drama, all they can do is leave us with heartbreaking silence. I heard the silence. It was deafening, and I rode with it. A funeral procession of empty air and hushed night. Still, not all was lost. In the mist I continued to jet on and off the road to the rhythm of passing cars, successfully grabbing exposed, living creatures, our eyes meeting and speaking, before I tossed them to safety. This of course slowed my progress, but what else could I do?

I arrived in Neosho when even the night was asleep. I found the address of a contact who'd given me permission to camp in their yard, and set up my tent in the shadowy protection of a tree tangled in a fence row. My plan had been to camp on the edge of town somewhere, but during my presentation earlier that evening the crowd had nearly mutinied with worry. One woman told me that I absolutely could not bike at night.

Typically I avoided such alarm by telling stories from my past and keeping my future plans secret. If my plans were discovered, my strategy was to listen to warnings, nod a lot, say "okay" a lot, then do whatever I had been planning. It always struck me as strange that people would try to convince me that what had been working for years wouldn't work another night. That doesn't mean I didn't heed warnings, but somehow

I could nearly always tell recommendations that came from experience apart from worry that came from assumptions, hype, and watching too much local news.

I looked at the woman who told me that I couldn't bike at night and said, "Yes, I can." It seemed like a ridiculous argument to be having, and one for which I certainly didn't have the energy. Danger was everywhere, and at night, with less traffic and my bright lights, her fears seemed overblown. Still, I felt bad contradicting her. Someone called a friend in Neosho and told them that I would be arriving that night to camp. A compromise.

What I feel bad saying, what I feel bad even thinking, is that every time I accepted help, I also accepted a slew of responsibilities. I had to use energy and time making plans, talking, explaining, worrying, following up, and keeping in touch. It was exhausting and sometimes I just wanted to camp and owe only the Earth. Don't get me wrong. Making connections was the whole purpose of my trip, and I wouldn't trade a single night I stayed with people for a camping night. Those connections made my trip possible. They were the stepping-stones I needed to get where I was going. I will be forever grateful. But it is also true that I wouldn't trade any night I camped alone in my tent for a night in a house with people. Those solitary nights also made my trip possible, and I will be forever grateful for them as well.

I settled into the gigantic lawn of the people whose names I had scribbled on some scratch paper. I burrowed into the cooling night, into my tent, into my sleeping bag, and into my thoughts. Like in a curtain call, I breathed a deep breath. As my chest rose, filling with the humid air filtered through the cut grass around me, I gathered the last few days into my lungs. Then I breathed out all I hoped to leave behind: my argumentativeness, my anger, my despair. I didn't know how to leave those things behind, but I knew I needed to try. I couldn't help the monarchs, or the planet, if I couldn't help myself. Another deep breath in, and out. I tried to let go of every death of every frog I hadn't been able to save.

My body settled into the stillness of another day done. Another day spent trying, failing, learning, and moving forward.

As I packed the next day, I brought with me the victories of the frogs whose lives I did save, could save. I carried their animal wisdom, strength, acceptance, and their silent protests. I also carried Val's dedication and the hope she nurtured with every community garden. I would remember, the next time I watched a butterfly sip nectar from a prairie slated to be a parking lot, that I was not alone.

I biked the short distance to my next school visit. In the staff bathroom I washed my face, donned my clean clothes, and took another deep breath as I stared at my reflection in the mirror. Watching myself watch myself, I pushed aside my despair and focused on trying. I walked to the gym and gave a voice to the monarchs. I proposed a better future to several hundred kids. "I haven't seen a monarch every day," I told the kids, "but every day I see the people that can help them. *We* can all help the monarchs."

We can all try. All we can do is try.

I left Neosho with only the outline of a plan: go south, make it to my presentations in Fayetteville, Arkansas, and figure out the rest along the way. I followed winding roads slick with rain through the growing hills of southern Missouri. Farms and forest mingled, and I stopped for each curiosity.

On a quarter-mile stretch of pavement, a herd of white grubs were making their way across the road. I squatted down, getting as close to eye level as I could to investigate. Each grub's six tiny legs, colored a beautiful rusty red, were raised to the sky, like puppies wanting their bellies rubbed. Instead of walking on their legs, they pulsed on their backs, moving steadily . . . somewhere. I apprehensively touched one, and the blob of white stopped scooting and curled in defense. I picked it up, along with around eighty of its fellow travelers, and tossed them back into the grass. They were June beetles. As larvae, they would spend their first year living underground, eating roots, annoying lawn owners, and growing big. Then, as shiny green beetles, they would emerge in June to fly awkwardly into our lives.

They were part of the team. They were turning lawns into homes, doing their part to reclaim the planet. They were bird food, frog food, and skunk food. They were beautiful in their own squishy way.

I hit Arkansas just before the 8000-mile mark of my trip. I noted my progress at the Welcome to Arkansas sign with my traditional untraditional photo shoot. I jumped, danced, kicked, and cheered alongside my tripod as cars buzzed by. I celebrated the miles I had covered, the monarchs I had met, and the changes in me—good and bad. How could I be both angrier and more hopeful than I was at the start of my trip? Instead of my emotions being diluted by time, they were adding up, intensifying.

Mile 8001: happier and sadder. Mile 8002: more energized and more tired.

The highway dropped me off at the top of the Regional Razorback Greenway bike trail, which wove through the conglomeration of northwest Arkansas's freeway cities, and eased the tensions of sprawl. The best miles of the trail were the miles fenced by thick stands of goldenrod. I watched the plumes of their yellow tentacles rise like waves, spill into the wind, and feed clusters of monarchs. I was always happy to see the goldenrod, like a good friend, waving to me. The pollinators must have been happy to see them, too. I have never met a lonely goldenrod.

My Fayetteville stop was a short one, and after one night and three presentations, I was back on the Razorback trail. I followed it to its end, then strayed east to visit one of my favorite places on the planet, a secret tucked into the woods and protected by stereotypes. Arkansas's Ozark Mountains are as wild, diverse, and adventurous as any place I have ever found. I felt them growing closer, even in the denuded hills covered in pastures and homes that framed them (and once were part of them). Pausing for the night in a lonely cemetery, I watched a monarch flutter overhead. I loved seeing monarchs in cemeteries. Whether or not you believe that the souls of our loved ones visit as monarchs, as Heather Schieder and I had pondered in Buffalo, the image of a monarch in a cemetery is beautiful. I longed to see each tombstone encircled not by cut grass but by wildflowers and their visitors. The beauty would be enough to bring the living and the dead to tears.

At the edge of the cemetery, where tombstones were more buried than not, the monarch and I set up camp. I unfolded my tent and the butterfly settled among the leaves of a branch. In the morning we would part ways.

The monarch would resume its trip, and I would enter the safety of public land and return home.

I had many homes; most were welcoming escapes into the protection of wild places. Watching the monarch fold herself into the swaying tree, I imagined we shared this sentiment. She was comfortable along highways and in towns, but at night, she was most at home in the wildest reaches.

I pedaled fast down the dirt roads, engulfed in Ozark forests. The unmarked turns were marked for me by memories. I breathed in my return to Arkansas's native lands, and rode to Tom and Cindy Rimkus's house, cradled in the bend of my favorite river.

If the monarchs hadn't had me on such a tight schedule, I would have spent a month at Tom and Cindy's, watching the vultures climb sky ladders to limestone bluffs, and crawling along dry streambeds carpeted with the crunchy leaves of summer's end. I would have visited the river, lingering as it dug into its rocky bed. And I would have chatted with my friends, who had carved a life in the woods and whose generosity and good humor were as reliable as the forest's quiet refuge. The monarchs had other plans. Their timeline permitted me two days.

I spent those two days recharging. Tom and Cindy's kindness and dedication, as well as their inspiring lives, lifted me up. I, too, wanted to shelter and be sheltered by a plot of land. I, too, wanted to live my life on my own terms. The river, the changing leaves in the trees, and the naked fish in shallow pools quieted my worried mind. They were my protection from the next inevitable round of bad news and political bullying. They were a reminder of why I was fighting.

Recharged, I made a presentation in the neighboring town of Huntsville, Arkansas, gave Tom and Cindy big hugs, and headed to my next appearance, in Austin, Texas, 700 miles south. I left with the gift of a tailwind and the company of clouds. The views were muted layers of forest. I let my memories paint them with a thousand greens.

As cars buzzed me and slaughtered roadkill stared unseeing at me, the good mood I had rediscovered at Tom and Cindy's seemed to evaporate. I knew then that I was in a rut. I had turned despondency into a habit. It felt reasonable—the product of toxic news, entitled drivers, and a long line of environmental injustices that I was constantly encountering up close and personal. The way I was responding was unhelpful and unhealthy, and I deliberately sought to change my perspective.

When two guys in a truck waved, I waved back. When they pulled over a mile later, I gathered my energy and goodwill and stopped. They were self-described rednecks, but of the progressive sort. They repeated everything I said with awe. "From *Mexico* . . . " they whispered, fascinated. The one with a bad eye volunteered a life lesson. He told me that after two of his girlfriends died, one after the other, he gave up drinking, drugs, and shitty food. He said, "I had to either get busy living or get busy dying."

I was definitely busy—but was I busy living?

When we parted, I took stock of my good fortune as a way to get busy living. I had a body I could tire out. I had enough money to not worry. I had confirmation that strangers were more often good than bad. I knew the despair, but I also knew the satisfaction of fighting it. I had the hope that we could come together and save the monarchs. I had found a rural road stretching like an arrow to the horizon and bursting with flowers hosting migrant visitors.

Down the road, another man, this time in a newer truck, slowed down and poked his head out the window. "You musta taken a wrong turn!" he said with a thick Arkansan accent.

I matched his smile, waved, and answered by continuing forward. No one could know for sure if they were on the right path, but one could have a feeling. I had a feeling, at that moment, that I was not lost; that I was where I needed to be, heading toward something bigger than myself. I was on this journey to understand and cope with my anger and sadness as much as I was to bike with butterflies. The monarchs brought beauty to their oppressors. They did not sink with grief. I would try to absorb this and their many other lessons.

I couldn't see Mexico, but I knew it was there. I couldn't see my future, but I knew I was on the right path. All I could do was stay busy living, getting lost, getting found, and moving forward.

The Southern Rewind

MILES 8157–8844

Since Kansas City, my route had become straighter, more direct. Like the monarchs, I was ready to get to Mexico. Unlike the monarchs, I would have been lost without my maps.

Just as the *pacing* of the migration seems tied to the sun, so too does monarch *navigation*. They seem to travel south in the fall and north in the spring using primary cues from the horizontal position of the sun in the sky. Since this position changes throughout the day, monarchs compensate for the daily shift using circadian clocks in their antennae.

On partially cloudy days, when the sun's position is obscured, the monarchs use polarized ultraviolet (UV) light as a secondary cue to navigate. Light from the sun becomes polarized as it travels through the atmosphere and is scattered by molecules. Since light is a spectrum of electromagnetic waves of varying sizes, different sized wavelengths scatter in predictable ways. Among the wavelengths that humans can see, known as the color spectrum, blue light is the shortest and most easily scattered. The result is a blue sky, though it is not a uniform blue. Light on the horizon has to pass through more atmosphere (and thus scatter more) than the light overhead, creating a gradient of color.

Monarchs can see the color spectrum as well as the UV spectrum. They can detect the pattern of UV light scattered across the sky, in a way similar to how we can see a gradation in the color blue. Though this pattern may be distorted with the presence of clouds, if there is a patch of blue sky, the monarchs can home in on its UV polarization patterns and continue to navigate in the correct direction.

When the sky is completely overcast, monarchs can't rely on the sun or UV polarization patterns to navigate, yet they can still manage. It seems that monarchs have a backup magnetic compass. Light sensitive crypto-chrome (CRY) proteins in their antennae react with even a limited amount of light in such a way as to trigger signals that allow monarchs to sense the magnetic field radiating from the north to south geomagnetic poles. The angle of this field, in relation to the surface of the earth, changes pre-dictably from the equator to the poles; the monarchs can sense this angle and navigate accordingly. The magnetic compass and the sun compass are slightly out of sync, so when the sun is out, the sun compass's trajectory overrides that of the magnetic compass.

Monarchs must also detect wind speed and direction to compensate for wind drift. Even though prevailing fall winds come out of the north-west, and can often aid migration, the butterflies still need mechanisms to adjust flight direction to compensate (just as we must do when we canoe across a lake on a windy day). The monarchs' antennae, able to sense smell, light, wind, vibrations, gravity, and atmospheric pressure, are likely key to dealing with this and other navigational challenges.

The more we know, the more we know just how much we don't know.

The more I learned, the more reverent I became.

On top of everything else, monarchs navigate south in the fall and north in the spring. It seems that during the winter, their internal compasses are recalibrated. This directional adjustment occurs after monarchs are exposed to three weeks of cold; conditions found at the overwintering grounds. Milkweed has a similar need, as seeds must undergo cold stratification to germinate. Such parallels, likely more than mere coincidence, should be added to the list of variables likely to be affected by climate change.

Though I could not see our planet as a butterfly does, I could bike down the highway and sense the monarchs' complicated but perfect wisdom. I could sense the layers invisible to my eyes and feel the elaborate nuances that weave together such brilliance. It's a feeling I imagine some describe as God's presence. When I entered Oklahoma, I was grateful that the sun was more than a star.

I crossed Oklahoma on a road better described as a paved scar. It stretched through the green-soaked, sunbaked forest. Four lanes wide, and still there was no space for me. A long, disobliging rumble strip gouged the shoulder. Unwelcomed, I was forced to ride the painted white line like a tightrope walker. Drivers didn't know why I wasn't retreating to the shoulder. They didn't notice the washboard ribs that made safer biking impossible. They didn't have the facts, but many apparently felt it was their job to teach me. Ignoring the three adjacent, empty lanes, many vehicles passed so close to me that I had to stop pedaling in order to brace against the wind they generated, wobbling to stay on a straight path. They were not teaching me anything. I already knew entitled drivers could be terrible.

It was the creatures nestled among the trees that extended a welcome. An armadillo, using its nose to plow the ditch alongside the highway, caught my eye and I froze. After seeing so many decaying, car-killed armadillo corpses, it was nice to observe one alive and well, if not exactly observant. I watched it dig and root and wander all the way to my feet. I made sure to not only look at the armadillo, but to *see* it. To see its rat-faced, turtle-shelled, pig-skinned, hawk-clawed cuteness.

The armadillo prodded my shoe, inspected my stationary presence, and discovered something worth scurrying toward—all part of its armadillo plan. When it scuttled onto the road, I followed it like a crossing guard. Though I was not powerful, I was not powerless. I could shoo armadillos into wooded safety.

I could also gather the stick insects wandering onto the pavement, inching their way from their camouflaged curtain of trees, to fling them back into their protection of pine needles. I tried not to fixate on the smashed bodies, the clear-cut forest of dead walking sticks, and instead looked for peace in the beauty of the still living.

The male walking sticks were like young saplings, green and feisty. They flailed their branch-like limbs more and wobbled faster than their browner, bigger, slower female counterparts. The female walking sticks didn't put up a fight, but instead would stretch their front legs out, to create an extension of their abdomen trunks. Sprawled out and frozen, they transformed into a stick, a branch, or a pine needle before my eyes.

A stick with eyes, an insect with bark. Nature thinks of everything.

My route only grazed the southeast corner of Oklahoma, and after two nights in the Sooner State, the wild summits of the Ouachita Mountains behind me, I crossed into Texas. With six days to cover the 345 miles to Austin, all of my time was given to the open road.

In one day I pedaled ninety miles through fields of bursting cotton and across scrublands abandoned by time. Behind fences, horses reached out expectantly. In the grassy ditch, a coiled rat snake sprang startled, as if he had wings. Another rat snake lay dying, the casualty of a callous car. I picked him up, looked him in the eye, and coiled all four feet of his gentle body in dappled light off the road. I continued as the sun set, the sky sketched with pencil lines of neon clouds. I camped next to a noisy air conditioner behind a church only slightly out of view.

Such days and their miles, not interrupted by presentations and meetings, blurred into a memory of blacktop and distant horizons.

I rode and then I rode some more. I rode far enough south that, like turning the dimmer switch on a light, I began seeing scissor-tailed flycatchers that were also migrating south and green milkweed that was now going to seed. They signaled a return to southern Texas. I was rewinding my trip, traveling back to the start where I was to end. I was 8500 miles from the start, and 1500 miles from the end, even though they were the same place.

Through Texas, alone with the road, standards took a back seat to convenience and practicality. My panniers were excellent refrigerators, but only when it was cold outside. In the Texas heat, food didn't last long. Meals like pasta were not reasonable, because by the time pasta, sauce, a few veggies, and some cheese were mixed together, they added up to leftovers that couldn't last. Instead I "cooked" lazy sandwiches. I'd take a bite of bread followed by consecutive bites of fixings—bell pepper, tomato, cheese. As my mom used to say when I worried about my applesauce touching my macaroni, "It all gets mixed up in your stomach."

Lazy sandwiches became breakfast, lunch, and dinner. In between I "prepared" lazy snacks. A bite each of apple, caramel baking chips, and peanuts made excellent caramel apples. A handful of cereal with a milk chaser was as good as a bowl of cereal, as long I drank the carton of milk in one day. Snack pairings never let me down. Bananas with peanut butter, apples with cheese, bell peppers with hummus. If carrots with pudding had sounded good, I could have made it happen. There were very few rules.

Camping spots were equally questionable. Two hundred miles from Austin, fences—tall enough to enclose the exotic deer people were giddy to shoot—sprang up. There was no way to get farther than thirty feet off the pavement, so I had to become invisible in plain sight. I took advantage of the highs and lows in the roads, bridges and ditches, juniper stands, tall grasses, and sharp curves to camp where camping didn't exist.

Each night I found some strange spot I could never have predicted, and each night the swirl of the sky entertained me. Fiery clouds spiraled toward the setting sun like genies being sucked into bottles; doodling clouds squiggled daily recaps. Once only darkness remained, my cover was secured and I always slept easy. Even if someone saw the moon's reflection on my bike, or my silhouette casting a strange shadow, who would stop?

There was one nod to civility I did not want to forego. Before I met Henry Gass, a reporter planning to bike with me into Austin, I was determined to have at least one shower in the preceding eight days. Lucky for me (and Henry), the bike touring gods came through for me in The Middle of Nowhere, Texas, the day before we met.

On a lonely stretch of highway, a sign announced the Falls on the Brazos. I turned onto the dirt road and followed its possibilities. The road ended and like a mirage, there sat a pavilion with working outlets and an unlocked bathhouse—with running water! There was no one to ask, no signs to forbid, so I accepted the miracle that had been bestowed. I plugged in my arsenal of electronics, showered, and finally, with the small bar of soap I carried, lathered my clothes until the suds turned black and swirled down the drain. I wasn't going for the "brand new" look, just one that was socially acceptable. When the rinse water was a dull gray, I wrung out my riding outfit and spread it along the chain lassoing the parking lot and swinging in the afternoon sun.

Cleaned, dried, charged, and refreshed, I gathered my gear and continued on. I wasn't subsisting, nor was I a nomad, but there was that quality to it. Even after years of finding what I needed on the road, my luck still amazed me. The world might not have been built for bike touring, but I liked transforming it into one that was.

There was just one more hurdle to cross before I met Henry the next morning: finding a place to camp. We had agreed to meet early the next morning in Taylor, Texas, so that we could bike to Austin together. Camping in towns was sketchy at best, but if I could find a shower when I needed one, I figured I could find a camping spot in Taylor.

The plan had been to figure it out when I got there. From fields of golden grasses spotted with green trees, to a haphazard collection of strip malls, the town of Taylor appeared. I made my way to a church to assess my camping options. While there were corners that were out of sight, they were already occupied by assorted trash, signs of someone else. I tried tucking behind several businesses that bordered plowed fields. Each attempt to sneak into the farmed void was met by watchful eyes. They were not threatening, but all the same I didn't want to reveal my spot so easily.

I took a chance, riding through apartment complexes toward the smokestacks of a factory and the brambles growing up around distant railroad tracks. I was uncertain of my aim but figured farther out of town was better than closer. Then, like every night I needed one, a camping spot revealed itself. At the edge of town, where the frontline of buildings marched into less developed areas, I spotted a social service building. It felt temporary, a modular structure rushed into service, wearing its age ungracefully. I stole behind it, finding a wide expanse of bare, tilled ground stubbled with corn stalks, partially tilled and haphazard. It was a Friday night, and I assumed no one would come to work on a Saturday morning. I could likely sleep peacefully. I organized my home as sunset began to spread across the sky. A silhouette of a rural town—electrical poles and arching cables, backyard trees and sloping roofs—lay under the arc of orange clouds highlighted by yellows, shadowed by purples, and set against the blues of a nearly done day. By the time it was dark, I was halfway through the pint of ice cream I had bought and kept insulated in my jacket.

I slept peacefully, the business's air conditioning unit humming like a distant ocean. In the morning, I brushed my teeth and rode back to the grocery store to meet Henry, where I was easy to spot. He placed a recorder in my pocket and off we went, into the wind, at a pace faster than my normal crawl, toward Austin. It was the longest stretch during which I'd had a biking companion, and it was nice to race after him rather than trying to keep up with my own shadow. The speed exhausted the last of my strength, however, and when we went our separate ways after reaching the city, I barely managed to limp to Nathan's house.

Nathan Nunn was my first host in Austin, but not the only one. For a week I flitted from house to house. From Nathan's, with a backyard for wildlife, I pedaled to Felicity and Paul Gatchell's home, where their adventurous lives were reflected in the art on their walls, the food on their table, and the stories they told. Next were two nights at Annie the Cat's house, where I pet-sat in exchange for my stay. Then it was to Elizabeth McBride's modern and cozy place. And finally, to Allison Jackson's, which was wonderfully full of hidden candies (fuel for my presentations to her students). Every place I stayed was its own mini adventure.

My work week in Austin started Sunday, with a public presentation at the Lady Bird Johnson Wildflower Center. The center is a cross between a native prairie and a botanical garden, where plants flourish, show off, and inspire. There were examples of native, themed, and ecotype gardens all woven together, accented by the dances of pollinators, the curiosity of visitors, and just enough signage to remind people that each plant had a name.

Each plant was visited by orbiting animal visitors, each flower the center of a miniature universe. The gulf fritillaries buzzed orange and silver as our sun played off their wings. The queen butterflies gathered together, looking alternately like monarchs or their own burgundy species, between wingbeats. The giant swallowtails hovered, and the occasional monarch darted through the flowers. So much potential was reflected in the eyes of every butterfly, the murmur of every bee, and the bloom of every blossom. Each made a silent request of the human visitors to transform their yards into habitat.

I walked the grounds in awe and felt the power and passion of Lady Bird Johnson, former first lady of our country and the founder and namesake of the center. Her legacy was one of beauty, casting light far beyond her time. I wanted to walk through the gardens with her and ask her how I could keep going. How I could harness some of her strength even when I felt weak. I imagined her soul blooming as flowers and sparkling as butterfly scales. All I could do was hold onto her example and keep trying in my own small way. I had to trust that just as a garden was made up of many flowers, saving the monarchs would be made up of many voices.

On Monday my work continued. Between school visits I looked up and saw the migration. The day was warm and the wind was weak, and I biked under a long line of monarchs. I counted about thirty of them, spaced apart like fluffy clouds, commuting south against the traffic below them. Each migrant rose and fell with its wingbeats, clearing the cars and adding grace to the city. Like falling leaves they floated through the sky, the autumn day performing a show that had been playing for millennia. Each butterfly was tracing an ancient path. The odds were stacked against them, yet there they were. I wanted to follow them, swap stories. Instead, I wished them luck. We both had jobs to do.

Over the next seven days I made sixteen appearances, telling the monarchs' story to over 1400 students in Austin. I was proof that adventurers, scientists, and conservationists were not superhuman. I just managed to reach out and follow a dream. Like seeing a monarch, talking with kids gave me strength and purpose.

As I packed up my tent after a presentation, a little girl walked up to me. At first she just watched as I managed to get everything back on my bike. Finally she asked her question. "Is any of this real?" she questioned, her brow furrowed with astonishment. "Yes," I told her, doing my best to match her sincerity.

Yes, I was doing something strange but real. Yes, she too could push boundaries and question the rules. Yes, the monarchs, their struggle, and the wisdom we could abandon were all real. It was our shared reality.

Tragically, I reflected, that little girl had not had a say in choosing the hardest reality—the declining health of the planet she'd inherited.

I tried to make my presentations to kids uplifting. I edited the doom, the gloom, and the heartbreak that had settled deep inside me. The pure truth would come, but I wanted to deliver a message of possibility. I wanted them to imagine all the skunks, owls, snakes, and butterflies they might see and hear and meet eye to eye. I wanted them to imagine all the gifts of ice cream, the forty-five-miles-per-hour downhill rides, and the epic camping spots they might find on their adventures. I wanted them to think about their part of the solution.

Even if they hadn't dug the hole, it was going to be their job to climb out.

The reality of this profound unfairness hit me deeply when I stayed with Elizabeth and her family, toward the end of my Tour d'Austin. Elizabeth's youngest son, Teddy, was nine years old, and he loved penguins. Holding his penguin stuffed animal, he told me about global climate change and what it was doing to the penguins and their Antarctic home. I gave Teddy a high five for thinking like a scientist and a steward, but inside I wanted to cry. I had learned about climate change in high school, and the reality of it had felt far away. Now, climate change was engulfing us and denial was futile. Teddy was a kid holding a stuffed animal in his arms and the dangers of climate change in his heart. He knew his penguins were in trouble.

We make younger generations carry our mistakes—our greed, selfishness, and entitlement. We make Teddy watch his penguins waddle toward extinction. Maybe Teddy will be able to hold the threats and find the answers. The least we can do for him and his cohort is to fight for their future. We can plant gardens and grow a future around them, full of life and hope and possibility.

In Austin, a coalition of teachers, students, parents, and community members have done just that, making space for schools to bloom with native plants. They are showcasing a better way, a future full of flowers, monarchs, and the beauty of learning by living. My favorite gardens were those shown to me by children as they pointed out flowers that butterflies had visited and leaves that caterpillars had eaten. Those young minds were likely to forget the facts they memorized for tests, but their gardens and the lessons cultivated would stick.

I waved goodbye to the students and their gardens, and rejoined the monarchs. Heading south, I felt free. I had the confidence of 9000 miles. I could practically hear the wings of the monarchs beginning to fill the oyamel firs. I had the goodwill of Austin and all the towns before it in my heart, the hope grown from school gardens in my soul, and the triumph of approaching the journey's end in my legs. Nothing could stop me.

A cold front, however, could slow me down and change my course.

Crossing
the Border

My plan had been to head toward Big Bend National Park, to see the Rio Grande wind a national border through desolation. I had been looking forward to the visit since planning my route, though as I got closer, the 500-mile detour through lonely desert had begun to feel burdensome. For the last few months, I had grown ambivalent about adding so many miles and diverging from the main migration path. It wasn't until I was three days out of Austin, sick for the first time on my trip, and freezing from a quickly approaching cold weather front that I decided to make a beeline (or butterfly-line) for Mexico.

By then I had returned to the Native American Seed Farm in Junction, Texas, where my northbound and southbound routes overlapped (a junction at Junction). On back roads I had retraced my spring route, this time with great effort. My throat tickled with pain, my head was heavy, and my body was weak. The previous day's sixty miles felt like 200. I curled up in the guest house with tea and willed myself to recuperate. The cold front lingered outside, forebodingly.

This time, the Neimans' farm felt more naked, like a fresh blanket of earth. But it was not bare. Some patches held newly sown seeds; others were stitched in the delicate green of small seedlings breaking the surface. Yet there were still areas of remnant color, where the farm's perennials grew, firmly rooted. Such a tapestry reminded me that this was no ordinary farm. Many Texas wildflowers, including those in rows cut by the farm's driveway, get their start in the fall, grow during the winter, and bloom in spring. While my trip was nearing its end, the stories of the Indian blankets, American basketflowers, and Texas yellowstars were just beginning. Beyond, on the wild fringes of the farm, there were more hints of fall: a mix of plants with hearty spines and browning greens. The sky, like a recurring dream, reset the clouds and lowered the sun sooner each day.

By deciding to save my Big Bend adventure for another time, I could instead focus on finishing my trip with the monarchs. As I gathered my strength, visited with my hosts—Bill, Jan, their daughter, Emily, and their new grandson, Fisher (who had been born between my visits, a reminder of how long I'd been following the monarchs)—I grew confident in my new plan. It was time to finish what I had started. It was time to return to Mexico.

I said goodbye, reversed course, and charged back into the wind. I rode through hills alive with junipers and paralleled the eons-old shores of the Frio River. I paused to jump into the river's cool waters. Basking on the shore, I watched a hawk leap from a tree with a lizard imprisoned in its talons and a dragonfly, glowing red, perch on a thorn before ambushing a passing insect. Back on the road, I rushed a gigantic caterpillar off the pavement and back into the roadside grasses where it disappeared with its camouflaged costume. I waved at monarchs hovering in the thorny branches of acacia trees as the sun set. I pedaled south until I caught sight of Mexico.

I arrived at the edge of Laredo, Texas, at night. It was a border town and its northern expanse was a veil of overpasses and highway signs. In a convenient swath of grass, below a forty-foot-tall terraced wall lifting up several lanes of traffic, I set up camp. Despite all the warnings, I felt safe. The twinkle of two cities' lights, worlds apart, met in space. My plan had been to wake up the next day and pedal into Mexico, but as I looked at the map and listened to the hum of traffic on the nearby road, a new idea emerged.

I decided I didn't want to take the main road, a four-lane superhighway, south into Mexico. The goal of the highway was to get as many people as possible through the 300-mile border gap as fast as possible. I knew this because I had already cycled that stretch once before. Now it presented itself more as a chore than an adventure. An alternative road paralleled the Rio Grande and the Mexican border. According to Google, it had a good shoulder. Crossing the border at Roma, Texas, a three-day ride south, looked much more interesting.

I had been told so many times to avoid the meandering swath that separates Mexico and the United States that I half expected nothing but snarling faces and machine guns along the way. The border carried so much drama, and that drama had me on guard. Yet each day, as the sun migrated between both worlds, I found only peace. The towns, the scenery, and the people felt just like the rest of my trip. The expansive blur of two countries felt like one.

What *was* pronounced on my ride along the border was not some boogeyman, but an ever-increasing crowd of colorful butterflies, not only monarchs, but a swirl of yellows, whites, silvers, oranges, blacks, reds, and in-betweens. I watched them traipse across the sky, as I questioned their destinations and stories. I could only assume their lives were as interesting and intricate as the monarchs' were. The whole world was worth saving.

For all the butterflies in the sky, there seemed to be an equal number of fading corpses on the gravel shoulder. I did my best to ignore the graveyard of colors and focus on the flurry alive above me. On a bridge, I scooped up a female monarch, flapping her wings futilely on the hot pavement. Several ants had already found her and I shook them off. I couldn't bear to let her die on boiling asphalt from the bites of unforgiving ants.

That monarch rode with me a few miles to my lunch spot. I gave her a half hour to fly away, but her wings were broken from being hit by a car. With every flap, she seemed to grow weaker. So close to her destination, her journey was to end at the border. With tears in my eyes, I told her that I would carry her spirit by bike to her destination. Then I cradled her in my hand, let the details of her decorated wings settle in my memory, and pinched her head and thorax. I had not killed her. An automobile had. Knowing this didn't make me feel better. I gave her back to the ants.

I was not the only person noticing the monarchs that were succumbing to traffic. Two independent teams of scientists surveyed short stretches of road across the monarchs' southern fall migration path. In Texas, one team found an *average* of 3.4 dead monarchs per 100-meter (328-foot) transect area in 2016 and 2017, with as many as 66 dead per transect. They extrapolated that an estimated 2–4 percent of the migrating population is killed in traffic collisions each year. Another team, surveying in high-concentration locations of Northern Mexico, encountered even more carnage. Along a 14-kilometer (8.9-mile) stretch of highway in Nuevo León, dead and injured monarchs were counted along 500-meter (1640-foot) transects. It was extrapolated that during the 2018 fall migration, in that stretch alone, 165,984 monarchs were killed. My heart was heavy imagining the weight of so many monarchs, their futures ripped from them—and us.

The mortally injured butterfly, now being eaten by ants in the grass, was the seventy-ninth monarch I had found dead or dying on my tour. I was noticing fewer than the scientists, in part because I was moving faster and with less focus. Drivers noticed even less. I was jealous, but understood that there was justice in acknowledgment. At least the deaths were being reported by the scientists. Their research was a scientific eulogy.

I carried on. My eyes and body returned, slowly, to the lives of the living. A cloud of butterflies swarmed like a storm around a flowering bush. Each rotation of my pedals was a renewed tribute to them. I had biked 9300 miles with, for, and because of them.

There was no wait to enter Mexico at the city of Ciudad Miguel Alemán; it was nothing but smiles and a sympathetic reintroduction to Spanish. I exchanged dollars for pesos, English for Spanish, miles for kilometers, and cell service for lots of questioning, pointing, and nodding. These shifts (and knowing the end was in sight) were the rituals I needed to reinvigorate myself. I had found that glorious blurry line between confusion and adventure.

My plan had been to get as far south as I could that same day. I took the smooth, quiet highway south, feeling smug and confident. Perfectly quiet miles bridged horizons that were sketched with scrubby shrubs, contemplative blue skies, and the peace of room to breathe.

Having cycled across the border between the United States and Mexico twice previously, I was well aware of the area's perceived dangers, thanks to the constant barrage of warnings. Warnings were not new to me, nor was meeting people who looked me in the eye and told me I was going to die. Biking on back roads in Bolivia, hiking across New Mexico, canoeing a reservoir in Montana, people frequently went beyond warning to inform me (with a mixture of pity and awe) that I *was* going to die. I learned to spot-sort panicked paranoia from first-hand experience. For me, worst case scenarios were just that.

That being said, when the man in the truck stopped twenty miles south of the border to tell me that fifteen men were waiting under a bridge with guns to kill me, I paused to mull this over. I noted that the man was not local, nor was he coming from the direction of said bridge. The drivers that had been coming from the suspected bridge had been waving, smiling, and cheering me on. I asked him how he knew, but I didn't understand his answer. Noting my skepticism, he raised his palms toward the sky, told me his hands were clean, and drove on, leaving me alone on the highway, weighing my options. I had no desire to meet fifteen bad guys with guns, but I didn't want to turn around either.

I decided I needed a second opinion. I waved down a truck coming toward me. I figured the woman driver and her older male passenger would have passed the supposed bridge and hopefully noted any potential danger. I told them what the other driver had said and asked their opinions. They laughed and told me I would be fine. Feeling slightly relieved, I ended the conversation with a joke I hoped translated to, "Thanks, take care, and if you see fifteen men with guns under a bridge, please don't tell them about me." They waved, and I took a deep breath. I would tell my parents about this in a decade. Maybe.

For the next ten miles I pounded across each bridge as if my life depended on it. My heart raced from the exercise and the threat. I was as alert as any prey could be. Nothing happened. The danger did not materialize. The bridges concealed only starved creeks.

The road forked several times, each an unpredictable divergence, and after twenty miles of vigilance, I knew I could relax. My haphazard route would have been impossible to guess; no one could be waiting for me. I could go back to worrying about real guns killing real people in real churches, schools, and concert venues in the United States. Fear must be considered in context. When the sun blinked tired, I turned off the road and camped between pointy shrubs and cow patties. It felt like the safest place on Earth.

Welcome Back

I didn't know where I was going to sleep on my second day back in Mexico. I needed water, but the road was long and lonesome. Nothing more than a handful of cars had passed me in several hours, and I had passed nothing more than a handful of cows. It was a relief when I finally spotted a house, shining in the light of the setting sun. Pulling up alongside the house's predictably makeshift store, I leaned my bike against the wall. I grabbed my empty bottles, because having them in hand always helped communicate what I was after.

Two kids were playing behind the counter, so I asked them about a way to fill up my bottles. Upon hearing my accented request, the kids, giggling, ran to collect more witnesses. In the resulting confusion, they led me to the porch of their home, where several women greeted me. They sat behind heaping piles of corn husks, which they were stuffing with *masa*; it was the tedious prep work for tamales. Before I could register anything more, I was placed behind a plate piled with food and shown where I could stay that night. After dinner, to show my appreciation, I scooted my chair into their circle and started to "help" stuff tamales. A gathering of men watched the spectacle, perhaps noticing for the first time that the work their wives, mothers, and sisters performed with ease was not as simple as it looked. The women already knew this, of course. Their laughter at my floundering skills put us all at ease.

While we worked, the youngest daughter bounced at my feet; first as a *coneja* (bunny), then as a *cangura* (kangaroo). The older children brought me butterflies they found sleeping among the potted plants, and dragged me from my toils with tamales to check out a frog showering under the drip of a leaky pipe. While the kids buzzed, the adults told me their stories. Some were light and cheerful, others were heavy with pain brought about by the imaginary line I had recently crossed with ease, thanks to my passport. Their family was split by governments and they had been forced to make impossible choices. Should they have stayed in the United States for their children, or were they right to have returned to care for their aging parents? All I could do was listen, hold their stories in my heart, and vow to cast my future votes with their lives in mind.

The butterflies the kids had found in the potted plants erupted the next day, and a kaleidoscope of wings unfolded in the sky. Never before had I seen such a rainbow of butterflies, and aside from the sanctuaries, I'd never seen such a concentration.

For a week I pedaled through a cloud of yellow sulphur butterflies that flew like they were learning to swim through air for the first time. Their yellow eyes matched their yellow wings, which held hints of orange, black, and white. Their numbers were not dense, like a mob passing in a panic. Instead, they were a calm, constant flow, spaced apart like stars moving across the sky; a synchronized pilgrimage of yellows.

While the sulphurs were the most abundant, it was the variety of butterflies that held me captive. As each species flew by, I tried to isolate it for a detailed study. The green and black stripes of long-winged butterflies were steady in the sky. The swallowtails with their dull greens and browns were leaflike but purposeful fliers. The large, bright orange fritillaries flew with flat, militaristic wing beats, looking highly efficient. Smaller fritillaries seemed to pause with each wing beat to show off their silvery undersides, which, like mirrors, reflected the sun.

Among the throngs, some of my favorites were the black-striped butterflies. Their stripes reminded me of horizons, each holding a setting sun. I took to calling the two distinct sizes "Big Sunsets" and "Little Sunsets." Another I was always happy to see was a large brown butterfly with wings dipped in orange and bisected by a white line that looked like it had been covered by painter's tape. Each time I observed their twilight colors, the beat of their brush-stroked wings, and their preference for dappled shade, I felt like I was meeting a friend of a friend.

Because of such variety in the winged fauna, including many big, orange butterflies, I was suspicious of monarch sightings. I needed to be close enough to see the clumsy, baggy gate or the folded wing keel of a monarch before committing to my identification. I needed to be close enough to see the yellow-orange shine through the veins of black, and confirm the absence of the burgundy hints and darker black fringe of the closely related queen butterfly. Seeing one or two monarchs a day kept me motivated. We were cheerleaders for each other, fellow stragglers, runners at the end of a marathon. Slowly but steadily, the finish line was creeping closer. We were not the first nor the fastest. We would get there when we got there.

I cheered on the butterflies all the way to Cuidad Victoria. There I aimed for the city's center to find a hotel and a shower, my first of both since Texas. Nestled between shops selling tortillas, bread, cell phones, popsicles, fruit, and meals made from every combination of tortillas, I found a modest hotel where the $10 a night price was somewhere between the cheapest dive and average. I ignored the hair in the shower drain and chewed gum on the nightstand, instead focusing on the Wi-Fi, the window connected to fresh air outside, and the nice employees who encouraged me to store my bike in my room. Once my bike was squeezed in, I needed to somersault over the bed to access the bathroom. Good enough.

Darlene Burgess from Point Pelee in Canada was tracking my progress, and when she saw where I was, she put me in touch with Iván Cumplián Medlin, a monarch steward, biologist, and outdoors enthusiast living in Ciudad Victoria. I biked to Iván's house to get some local intel on route options, and wound up staying in the city an extra day to enjoy his company.

Iván worked for the government studying monarchs and encouraging tourism in his state of Tamaulipas, where each fall, fields of flowers brimmed with migrating monarchs stockpiling fat for the winter. Knowing that these monarch hotspots were as important as the overwintering forest farther south, the government was investing in programs that would encourage farmers to leave their fields of flowers untouched. A helpful bounty, reserved for the winged migrants.

Iván also worked with a tourist company, taking visitors canyoneering and hiking. He showed me photos of trips he'd led through milky green waters, the obvious and beautiful result of limestone filtration. When I remarked on the color and mentioned limestone, he confirmed my observation. He also confirmed what I had suspected, that he saw Mexico differently than most: he saw its natural gifts with deep appreciation. In a world where most look at snakes, frogs, butterflies, limestone, and long steep hills, but don't see the beauty of these things, it was comforting to meet a kindred spirit.

Iván was also my age. I'd mostly been working with people older than I was, or students. It had been a long ten months, and I was smitten with him immediately. I followed his recommended route out of town, half because it sounded like a good idea and half because I was powerless in the presence of his charm.

The road Iván pointed me toward climbed slowly out of Ciudad Victoria and its collection of smog, noise, trash, lights, and traffic. As I climbed, I was once again accompanied by butterflies. Together we gained ground and perspective. Distant mountains sprouted on the horizon and the valley shrank away. Just when I was tired enough to no longer be enjoying the climb, I reached the summit. There, the bluest lizard I have ever seen—make that the bluest *animal* I have ever seen—welcomed me. He wore a golden vest and a black necklace, but these were only accessories. His neon blue scales were the showstopper. Even when I finally rocketed down the other side of the mountain, into the hot, waiting desert, that blue remained vivid in my mind. It still does. It reminds me that a forest is not green. A desert is not brown. The details are sprinkles of every color. Leaning into a curve, I dodged a jet-black garter snake basking in the radiant heat of the pavement. Backtracking to move her off the road, I noticed her pixelated orange racing stripe, shimmering like jewels.

Below, the desert spread unmeasurable, and the road twisted like the bowels of a thirsty beast. The distant balding mountains held a million hues of olive, emerald, and blue, and kept time in the wrinkles of their eroded canyons. Yuccas poked up like jagged brushstrokes of life. I traced a canyon cut by desert floods, stopped in a town run by a mob of macaws, gave my best shot at a presentation in Spanish, and returned to Tula, Tamaulipas, for the second time on my trip.

Instead of being met by a man on a motorcycle offering me free ice cream, I was met by Elda Margarita Villasana Rojas, Carlos Adrián Verdin Licon, and their two kids, all cyclists. We had met nearly eight months earlier on the road, and they had extended an invitation to visit on my return. My plan was to stay two days at their house. I ended up staying five. Five days to play Scrabble (in Spanish, with a three-year-old team partner—and we won), make *gorditas* and clear tables at their restaurant, and ride bikes to a local pyramid. I got to cheer on Diego, the youngest son, as he marched in a parade, and look for tadpoles with Carlos, the oldest son. My stay was documented by the kids, who used my camera to take photos—mostly of the backyard chickens, pig, and goats. It was a tornado of life that left me exhausted, but with untradeable memories.

"Voy a hacerte cosquillas," Diego announced as he ran toward me.

"What are cosquillas?" I asked.

He answered by putting his hands up and wriggling his fingers playfully. *Cosquillas*, I soon learned, are tickles.

Leaving was not easy.

The Homestretch

MILES 9693–10,201

The final stretch, less than 400 miles, spanned before me, and memories of the start flickered in my mind. I was not looking forward to reuniting with the steep hills and traffic-choked roads of my early miles. I hoped my parallel route among the fertile, green, butterfly-crowded slopes of the Sierra Madre mountain range would be easier.

For two days I managed a fast pace on relatively flat ground. Fields of sugarcane surrounded me. Mile after mile, I pieced together the steps of the sugar-making process. I passed fields that were once green forests, now filled with sweet stalks grown in formation. I passed men with machetes cutting down the ready shoots and balancing them in piles. Trucks overflowing with the harvest raced down the road to processing plants identified by belching smokestacks. I passed fields capped in plumes of black smoke that filled the sky with a dusty glow, shielding the bordering mountains from my view.

The road hinted at the mountains that closed in, but remained hesitant, maintaining its balance without dramatic climbs or drastic descents. When the farm traffic fell away I was left alone, to pass only the bravest of family farms clinging to the steep hillsides next to me. The topography meant camping was out of the question. When darkness won its nightly battle, I entered a town, hoping to find a fire station.

On my other bike trip through Latin America, fire stations had proved to be great places to camp, but this town didn't have one. At the tourist office in the town plaza I explained my predicament, and a man signaled for me to follow him. We walked across town, my bike and my mission the fodder for small talk. He likely didn't understand all I said. I certainly misunderstood where he was leading me. We arrived, not at a field where I could safely pitch my tent, but instead at a small building. His knock was answered by a man who shrugged at the request and led me inside. There were showers, bathrooms, a kitchen, and two rooms full of bunk beds. I was told which room was for women, and without further ado I found myself confused, sheepish, and apparently at home.

I made a caldron's worth of sugary coffee with some of the tenants and the watchman, and put the pieces together: I was in a bunkhouse for people who lived in the countryside and needed a safe, affordable (free) place to stay when they came to town for doctor visits or to sell their goods. I offered a donation to the watchman, who said it was unnecessary.

I barely slept, as a two-week-old baby cried to her mother a few bunks over. I tossed and turned, but was content. Another memorable adventure. Five-star hotels were forgettable. That bunkhouse was not.

Poorly rested the next morning, I continued down the narrow back roads before intersecting with a main highway, which, unlike the previous day's road, climbed maniacally up and over the steep mountain ridge. There was no racing to the top; it was a slow, steady, all-day climb. When it was getting to be time to camp, I looked to my right and saw the wall of rock the road had been cut from. To my left was a near cliff. The sun grew weak and the racoon-like coatimundis scampered into the thick forest. I would once again have to get creative about camping.

After talking with officials in several government buildings in whatever town I found myself in at nightfall, and rejecting one spot because it was flooded with lights and another because it was a crowded pathway, I finally found a secluded soccer field. I was told not to camp in the dark shadows, and though I would have felt safer there, I listened because I was their guest. My tent went up near a house with an anxious dog; he left me alone after a few well-aimed (but merciful) rocks explained the situation. The temperature plummeted that night. I burrowed into my sleeping bag. If hunger makes the best seasoning, then exhaustion makes the fluffiest bed.

The mountains constricted the roads and my options, but I still managed to find a less-traveled way forward. Near a summit I chose not to fly down the other side on the main road, but to cut across the variegated ridge, adding 6000 feet of extra climbing toward another thoroughfare. It was slow riding, but the twists and turns, the dirt and potholes, the vibrant forest flecked with butterflies and flowers, and the views of seemingly endless mountains gave me strength. I was not daydreaming about the end, I was happy to be exactly where I was. As good as my route seemed, it was even better after I stopped at a restaurant for some beans and rice.

After crossing into Mexico, I had settled into a routine of buying lunch or early dinner at a restaurant once a day. Each meal cost less than $3 and left me so full I didn't need to eat anything other than a banana or mandarin orange until the next day. The stops were also good opportunities to fill up my water bottles and charge my phone.

Finishing up my beans and rice, so full I was on the brink of pain, I started a conversation about the road ahead to stall getting back on my bike. Soon several groups had joined the discussion, including Chloe, who told me that I had only one option: to stay at her house. I agreed.

Chloe was a strong, spirited 89-year-old woman who could have passed for 65. She lived in a house above the highway, but had a spare one-room house below the highway, where I could stay. Though the spare house was small, its window overlooking the valley was huge. I sat on one of the two beds and studied the steep green terrain, decorated with the houses of her small village. It was perhaps my favorite homestay view of the entire trip. The not-so-hidden bathroom was on the outside deck and made good use of the panorama.

If my timing or my incidental choices of where to stop and eat had been different, this small but meaningful episode would have been lost. Traveling made it easy to see the fabric of fate, the brilliance of planning on the unplannable.

In the morning I showered at Chloe's house. Most people living in the countryside in Mexico don't have showers per se. They heat water on the stove, mix it with cooler water in a bucket, and use a bowl to throw water over themselves in a small bathroom with cement floors and a drain. I liked the ceremony it created, and found myself appreciative of warm water poured cup by cup. When I emerged clean, Chloe made certain, with a hilarious and grandmotherly pantomime, that I had washed *everywhere*. Another memory. Another gift.

I left Chloe's house with less than 250 miles to go. Two hundred and fifty miles of seemingly endless climbs, broken only by their invigorating downhill equivalents. Many of the miles blurred as I threw myself into an audiobook, an attempt to ignore the pollution, noise, and constant stares of heavy traffic. The downhill rides regathered my focus and stand out in my memory. At forty-five miles per hour, the sorry guard rail and sliver of pavement blurred until I felt as if there was nothing between me and a thousand-foot fall. Staying upright demanded concentration, and I could only absorb the pure thrill of speed, wind, and recklessness. It made me feel strong, like an animal.

The sanctuaries got closer. Traffic ebbed and flowed as I picked my route along busy highways and dirt-path back roads. I got lost in a labyrinth of cobblestone. I found my way with directions from an eight-year-old boy. One night, I hurtled myself and my bike over a fence to camp under the safety of a heavy limbed tree. One day I ate too much pig-shaped bread, because I had missed it. The next day, I began to climb the last paved road to the sanctuaries.

As I climbed, the traffic abated, the road rose above the smog, the hills were reclaimed by forest, and peace began to find me. I climbed for an hour, contemplating the thousands of miles that had led me to where I was. Every pain, every doubt, every decision could finally drain away. They did not need my energy. I was a day away from finishing.

Just short of the summit, I rejoined my route. Nine and a half months earlier I had come to the same junction and made my first wrong turn. I had stood there wondering where to go, which path to take, what route was correct. I had gone left, a supposed mistake. But every mistake had added up to my arrival right where I was supposed to be. I'll never know what would have happened had I not made that left turn. I owed my trip to that so-called mistake.

I left my shadow contemplating the road. The last thirty miles dove up and down, mirroring the first thirty miles. I no longer needed a map. I knew where I was, and I knew where I was going. I traced the hill and reentered Michoacán for the second time in 10,170 miles. At the pass was a flat meadow, which I crossed before beginning the last downhill stretch. On hairpin turns I passed trucks, confident that there were no speed bumps until the straightaway that dropped into Angangueo. The forest puddled in my vision, thanks to eyes made watery by the wind and my speed.

I paused in Angangueo long enough to stuff myself with *tortas*, before saddling up for the hardest climb of the trip. There was nothing left to do but climb. Nothing left to do but finish.

From afar, the road—one of the steepest I've ever tried to ascend on a bike—looked smooth enough. Up close, I could feel the jar and jostle of each gap, bump, and crater in the cobblestones. I felt like a pitiful cowboy, graceless and tired, riding a hobbled bronco.

I leaned into my bicycle and summoned every muscle in my body to creep forward, the biking equivalent of limping. A few kids looked up from giggling and chasing a ball to follow me with their eyes. The turn of their heads matched my agonizing pace. A man with a chicken under his arm stared at me, trying to figure out where I had started.

"Dónde viene?" he questioned as I went by.

"Allá," I grinned, nodding toward the top of the steep hill as I pedaled. Up there was where I had started. Up there was where I was headed.

I felt like the hill was asking me, How much do you want this? I responded by rising from my bike saddle and pitching my weight back and forth. My answer was, A lot. At this point, no hill could stop me. I was going to make it to the top, and become the first person to follow the migrating monarch butterflies by bicycle on their entire migratory loop.

Out of town and away from the spectators and their bemused looks, I was grateful not to have an audience watching my struggle. I could turn my attention to the mountain views, a once-uninterrupted forest now quilted with fields, pastures, and homes.

A monarch butterfly wandered into view. It was a welcome interruption and I stopped to watch her pass effortlessly overhead, her stained-glass wings catching sunlight while I caught my breath. As my muscles reveled in the pause, the monarch's modest passing proved once again what I had long ago learned: monarchs are tougher than cyclists. I traced the road and gauged our progress. "Almost there," I told her.

Ten thousand two hundred miles down. One mile to go.

The monarch bobbed out of sight. She would easily beat me to her overwintering grounds. Instinct would pull her to the trees her great-grandparents had settled in the winter before. She was returning home, even though she had never been there before. I was returning, too. One mile shy of connecting the start with the end, spring with fall, descendants with their ancestors, the first page with the last. By then, however, I knew that my trip would never really end. Even in a mile, when my bike ride was done, the monarchs would forever flutter in my heart.

I was not surprised that tears came. Tears created not by wind or pain, not even because I had nearly accomplished a dream. The tears were for that monarch. I was watching her complete a wondrous journey. An odyssey defined by science and God. I was part of something beautiful, connecting beautiful things. Was it possible that her great-grandparents had floated in the sky alongside my bicycle? That she was a caterpillar I had visited at a school, or the one I had saved from the mower?

I proceeded with caution. I had thousands of days' worth of adventures, but only a few endings. I knew that at the conclusion of a long journey there would be a collision of joy and loss, satisfaction and bewilderment. I was grateful for my past finales, because they had taught me to prepare for the end. Without the daily marking of miles and the constant goals of finding food and shelter, finishing a trip could leave me unsettled, adrift. Shadows could hide in the happiness of accomplishment. Knowing this, I was prepared as I crested the last wave of road and the parking lot of El Rosario appeared like a mecca before me. Both charged and hesitant, I rode toward the finish line.

At the ticket booth, a familiar face greeted me. He told me Brianda was waiting for me just up the hill, then returned my money. I smiled a smile I couldn't tame, and found Brianda waiting. Hugs, handshakes, drinks, and treats ensued. I was showered with tokens of congratulations. I focused on the sweet cold of a soda, the smiles of my friends, questions I could barely answer, and the nearly meditative state of not worrying about the days to come. But my trip was not yet over.

Two hundred and fifty-five days after starting, I parked my bike and completed the last leg of my roundtrip journey on foot. Walking with my head tilted back, I caught sight of the colony above. A kaleidoscope of monarchs saturated the hushed forest in mute color. My gaze climbed the trees, which were bent with the weight of my traveling companions. There were millions of them.

We had arrived.

The monarchs' huddled bodies blended into the period of a seemingly impossible sentence. They exist: these tiny creatures capable of navigating a continent's worth of impossible odds. And while the hazards they face may multiply, so too grows their legion of defenders. The defenders know that together, we—butterflies, humans, and our collective neighborhood of creatures—can be great.

Together, many monarchs render a migration. Together, many miles create an adventure. Together, many gardens produce solutions. Our collective voices add up to change.

As long as there are adventure takers, solution makers, change instigators, and migrants, our actions together equal hope.

I kept my gaze on the trees, but closed my eyes just for a moment, to breathe the silence. There was so much to say and at the same time nothing to say at all. This bike ride was over. But there was still so much work left to do. I, you, we must keep moving forward. Even if it's just one slow mile at a time.

Selected References

ARRIVING AT THE START

Anderson, J. B., and L. P. Brower. 1996. "Freeze-Protection of Overwintering Monarch Butterflies in Mexico: Critical Role of the Forest as a Blanket and an Umbrella." *Ecological Entomology* 21: 107–116. https://doi.org/10.1111/j.1365-2311.1996.tb01177.x

Brower, L. P., E. H. Williams, D. A. Slayback, L. S. Fink, M. I. Ramírez, R. R. Zubieta, M. Ivan Limon Garcia, P. Gier, J. A. Lear, and T. Van Hook. 2009. "Oyamel Fir Forest Trunks Provide Thermal Advantages for Overwintering Monarch Butterflies in Mexico." *Insect Conservation and Diversity* 2: 163–175. https://doi.org/10.1111/j.1752-4598.2009.00052.x

Oberhauser, K. S., and A. Peterson. 2003. "Modeling Current and Future Potential Wintering Distributions of Eastern North American Monarch Butterflies." *Proceedings of the National Academy of Sciences of the United States of America* 100 (24): 14063–14068. http://www.jstor.org/stable/3148912

Rendón-Salinas, E., and G. Tavera-Alonso. 2013. "Monitoreo de la Superficie Forestal Ocupada por las Colonias de Hibernación de la Mariposa Monarca en Diciembre de 2013." https://www.telcel.com/content/dam/telcelcom/mundo-telcel/sala-prensa/noticias/archivos/2014/enero/monitoreo-monarca.pdf

Rendón-Salinas, E., F. Martínez-Meza, M. Mendoza-Pérez, M. Cruz-Piña, G. Mondragon-Contreras, and A. Martínez-Pacheco. 2019. "Superficie Forestal Ocupada por las Colonias de Hibernación de Mariposa Monarca en Mexico Durante la Hibernación de 2018–2019." https://d2ouvy59p0dg6k.cloudfront.net/downloads/2018_reporte_monitoreo_mariposa_monarca_mexico_2018_2019.pdf

United States Census Bureau. https://www.census.gov.

THE MONARCHS' WINTER NEIGHBORS

Brower, L.P., D. R. Kust, E. Rendón-Salinas, E. García-Serrano, K. R. Kust, J. Miller, C. Fernandez Del Rey, and K. Pape. 2004. "Catastrophic Winter Storm Mortality of Monarch Butterflies in Mexico in January 2002." In *The Monarch Butterfly: Biology and Conservation*, edited by K. M. Oberhauser and M. J. Solensky. Ithaca, NY: Cornell University Press. 151–166.

Eligio García Serrano, El Fondo Monarca. Personal communication with author, June 7, 2020.

Missrie, M. 2004. "Design and Implementation of a New Protected Area for Overwintering Monarch Butterflies in Mexico." In *The Monarch Butterfly: Biology and Conservation*, edited by K. S. Oberhauser and M. J. Solensky. Ithaca, NY: Cornell University Press." 141–150.

Monarch Butterfly Biosphere Reserve World Heritage Site nomination document. 2007. https://whc.unesco.org/uploads/nominations/1290.pdf

Mónica Missrie, Monarch Butterfly Fund. Personal communication with author, June 8, 2020.

O. R. Taylor, Jr. Personal communication with author, June 7, 2020.

Savko, M. S. 2002. "Ejidos, Monarchs, and Sustainability: Forest Management and Conservation in the Monarch Butterfly Biosphere Reserve of Mexico." Thesis. Oregon State University. https://ir.library.oregonstate.edu/concern/undergraduate_thesis_or_projects/x059cd312

Tucker, C. M. 2004. "Community Institutions and Forest Management in Mexico's Monarch Butterfly Reserve." *Society & Natural Resources* 17 (7): 569–587. http://dx.doi.org/10.1080/08941920490466143

Vidal O., J. López-García, and E. Rendón-Salinas. 2013. "Trends in Deforestation and Forest Degradation after a Decade of Monitoring in the Monarch Butterfly Biosphere Reserve in Mexico." *Conservation Biology* 28 (1): 177–186. https://doi.org/10.1111/cobi.12138

DESERTED MILES AND TRIALS

Alonso-Mejía, A., E. Rendón-Salinas, E. Montesinos-Patiño, and L. Brower. 1997. "Use of Lipid Reserves by Monarch Butterflies Overwintering in Mexico: Implications for Conservation." *Ecological Applications* 7 (3): 934–947. https://doi.org/10.1890/1051-0761(1997)007[0934:UOLRBM]2.0.CO;2

Arellano-Guillermo, A., J. I. Glendinning, and L. P. Brower. 1990. "Interspecific Comparisons of the Foraging Dynamics of Black-Backed Orioles and Blackheaded Grosbeaks on Overwintering Monarch Butterflies in Mexico." In *Biology and Conservation of the Monarch Butterfly*, edited by S. B. Malcolm and M. P. Zalucki. Los Angeles: Los Angeles County Natural History Museum. 315–22.

Brower, L. P., C. J. Nelson, J. N. Seiber, L. S. Fink, and C. Bond. 1988. "Exaptation as an Alternative to Coevolution in the Cardenolide-Based Chemical Defense of Monarch Butterflies (*Danaus plexippus L.*) Against Avian Predators." In *Chemical Mediation of Coevolution*, edited by K. C. Spencer. New York: Academic Press. 447—475.

Journey North. https://journeynorth.org/tm/monarch/sl/4/who_is_eating_monarchs.pdf

Journey North. https://journeynorth.org/tm/monarch/sl/20/TG.html

Monarch Joint Venture. https://monarchjointventure.org/news-events/news/fall-migration-how-do-they-do-it

Monarch Watch. https://monarchwatch.org/biology/pred2.htm

Oberhauser, K. S. 2004. "Overview of Monarch Breeding Biology." In *The Monarch Butterfly: Biology and Conservation*, edited by K. M. Oberhauser and M. J. Solensky. Ithaca, NY: Cornell University Press.

A MILKWEED GREETING

Baumle, K. 2017. *The Monarch: Saving Our Most-Loved Butterfly*. Pittsburgh, PA: St. Lynn's Press.

Frey, D. 1997. "Resistance to Mating by Female Monarch Butterflies." In *1997 North American Conference on the Monarch Butterfly*, edited by J. Hoth, L. Merino, K. Oberhauser, I. Pisanty, S. Price, and T. Wilkinson. Montreal: Commission for Environmental Cooperation. 79–87.

Hill, H. F., A. M. Wenner, and P. H. Wells. 1976. "Reproductive Behavior in an Overwintering Aggregation of Monarch Butterflies." *The American Midland Naturalist* 95 (1): 10–19.

Journey North. https://journeynorth.org/tm/monarch/LarvaFacts.html

Journey North. https://maps.journeynorth.org/maps

Monarch Joint Venture. https://monarchjointventure.org/monarch-biology/life-cycle/larva/guide-to-monarch-instars

Monarch Watch. https://monarchwatch.org/press/press-briefing.html

O. R. Taylor, Jr. Personal communication with author, December 20, 2019.

Oberhauser, K. S. 2004. "Overview of Monarch Breeding Biology." In *The Monarch Butterfly: Biology and Conservation*, edited by K. S. Oberhauser and M. J. Solensky. Ithaca, NY: Cornell University Press.

Oberhauser K. S., and D. Frey. 1999. "Coercive Mating by Overwintering Male Monarch Butterflies." In *The 1997 North American Conference on the Monarch Butterfly*, edited by J. Hoth, L. Merino, K. Oberhauser, I. Pisanty, S. Price, and T. Wilkinson. Montreal: Commission for Environmental Cooperation. 67–78.

Solensky, M. J., and K. S. Oberhauser. 2004. "Behavioral and Genetic Components of Male Mating Success in Monarchs." In *The Monarch Butterfly: Biology and Conservation*, edited by K. M. Oberhauser and M. J. Solensky. Ithaca, NY: Cornell University Press.

FINDING REFUGE

Journey North. https://journeynorth.org/monarchs/resources/article/facts-monarch-butterfly-ecology

Monarch Watch. https://monarchwatch.org/press/press-briefing.html

CHASING SPRING

Fadiman, D. "Lincoln Brower on What Good is a Butterfly" (video). McKenzie, Marlo (producer/editor) Vimeo. https://vimeo.com/123251908

Monarch Watch. https://monarchwatch.org/blog/2020/06/03/monarch-annual-cycle-migrations-and-the-number-of-generations/

Monarch Watch. https://monarchwatch.org/press/press-briefing.html

O. R. Taylor, Jr. Personal communication with author, June 15, 2019.

Oberhauser, K. S., A. Alonso, S. B. Malcolm, E. H. Williams, and M. P. Zalucki. 2019. "Lincoln Brower, Champion for Monarchs." *Frontiers in Ecology and Evolution* 7 (149). https://doi.org/10.3389/fevo.2019.00149

United States Fish and Wildlife Service. "Petition to Protect the Monarch Butterfly (*Danaus plexippus*) Under the Endangered Species Act." http://ecos.fws.gov/docs/petitions/92210//730.pdf

Vidal, O., and E. Rendón-Salinas. 2014. "Dynamics and Trends of Overwintering Colonies of the Monarch Butterfly in Mexico." *Biological Conservation* (180): 165–175. https://doi.org/10.1016/j.biocon.2014.09.041

REMEMBERING TALLGRASS

Brower, L. P. 1997. "Biological Necessities for Monarch Butterfly Overwintering in Relation to the Oyamel Forest Ecosystem in Mexico." In *The 1997 North American Conference on the Monarch Butterfly*, edited by J. Hoth, L. Merino, K. Oberhauser, I. Pisanty, S. Price, T. Wilkinson. Montreal: Commission for Environmental Cooperation. 11–28.

Calvert, B. 2004. "Two Methods of Estimating Overwintering Monarch Population in Mexico." In *The Monarch Butterfly: Biology and Conservation*, edited by K. S. Oberhauser and M. J. Solensky. Ithaca, NY: Cornell University Press.

Craddock, H. A., D. Huang, P. C. Turner, L. Quirós-Alcalé, and D. C. Payne-Sturges. 2019. "Trends in Neonicotinoid Pesticide Residues in Food and Water in the United States, 1999–2015." *Environmental Health* 18 (7). https://doi.org/10.1186/s12940-018-0441-7

Hartzler, R. G., and D. D. Buhler. 2000. "Occurrence of Common Milkweed (*Asclepias syriaca*) in Cropland and Adjacent Areas." *Agronomy Publications* 32. https://lib.dr.iastate.edu/agron_pubs/32

Kansas State University. https://www.k-state.edu/seek/spring2017/konza/index.html

Monarch Joint Venture. https://monarchjointventure.org/news-events/news/measuring-monarchs-and-milkweed

Monarch Joint Venture. https://monarchjointventure.org/images/uploads/documents/risks_of_neonics_to_pollinators.pdf

Monarch Watch. https://monarchwatch.org/blog/2020/03/13/monarch-population-status-42/

National Park System. https://www.nps.gov/tapr/learn/nature/a-complex-prairie-ecosystem.htm

Rendón-Salinas, E., and G. Tavera-Alonso. 2013. "*Monitoreo de la Superficie Forestal Ocupada por las Colonias de Hibernación de la Mariposa Monarca en Diciembre de 2013.*" https://www.telcel.com/content/dam/telcelcom/mundo-telcel/sala-prensa/noticias/archivos/2014/enero/monitoreo-monarca.pdf

Rendón-Salinas, E., F. Martínez-Meza, M. Mendoza-Pérez, M. Cruz-Piña, G. Mondragon-Contreras, and A. Martínez-Pacheco. 2019. "*Superficie Forestal Ocupada por las Colonias de Hibernación de Mariposa Monarca en Mexico Durante la Hibernación de 2018-2019.*" https://d2ouvy59p0dg6k.cloudfront.net/downloads/2018_reporte_monitoreo_mariposa_monarca_mexico_2018_2019.pdf

The Nature Conservancy. 2010. Tallgrass prairie comparison map. "Home of the Range." *Seek* 7 (1). https://newprairiepress.org/cgi/viewcontent.cgi?article=1178&context=seek

Thogmartin, W. E., J. E. Diffendorfer, L. López-Hoffman, K. Oberhauser, J. Pleasants, B. X. Semmens, D. Semmenss, O. R. Taylor Jr., R. Wiederholt. 2017. "Density Estimates of Monarch Butterflies Overwintering in Central Mexico." *PeerJ* 5:e3221. https://doi.org/10.7717/peerj.3221

Vidal O., J. López-García, and E. Rendón-Salinas 2014. "Trends in Deforestation and Forest Degradation After a Decade of Monitoring in the Monarch Butterfly Biosphere Reserve in Mexico." *Conservation Biology* 28 (1):177–86. https://doi.org/10.1111/cobi.12138

Zaya, D. N., I. S. Pearse, G. Spyreas. 2017. "Long-Term Trends in Midwestern Milkweed Abundances and Their Relevance to Monarch Butterfly Declines." BioScience 67 (4): 343–356. https://doi.org/10.1093/biosci/biw186

HARNESSING SCIENCE

Monarch Watch. https://monarchwatch.org/waystations/

O. R. Taylor, Jr. Personal communication with author, December 27, 2019.

Taylor, O. R. Jr., J. P. Lovett, D. L. Gibo, E. L. Weiser, W. E. Thogmartin, D. J. Semmens, J. E. Diffendorfer, J. M. Pleasants, S. D. Pecoraro, and R. Grundel. 2019. "Is the Timing, Pace, and Success of the Monarch Migration Associated with Sun Angle?" *Frontiers in Ecology and Evolution* 7: 442. https://doi.org/10.3389/fevo.2019.00442

HOPE IN THE CORN

Capehart, T., and S. Proper. 2019. "Corn is America's Largest Crop in 2019." *USDA Economic Research Services* blog. https://www.usda.gov/media/blog/2019/07/29/corn-americas-largest-crop-2019

Clark, P. 2012. "Milkweed Fruits: Pods of Plenty." *The Washington Post: Urban Jungle*, September 25. https://www.washingtonpost.com/wp-srv/special/metro/urban-jungle/pages/120925.html

Davis, A. K., H. Schroeder, I. Yeager, and J. Pearce. 2018. "Effects of Simulated Highway Noise on Heart Rates of Larval Monarch Butterflies, *Danaus Plexippus*: Implications for Roadside Habitat Suitability." *Biology Letters* 14 (5). http://dx.doi.org/10.1098/rsbl.2018.0018

Emilie Snell-Rood. Personal communication with author, December 17, 2019.

Hollingsworth, J. 2019. "Climate Change Could Pose 'Existential Threat' by 2050." *CNN*, June 4. https://www.cnn.com/2019/06/04/health/climate-change-existential-threat-report-intl/index.html

Pimentel, D. 2003. "Ethanol Fuels: Energy Balance, Economics, and Environmental Impacts Are Negative." *Natural Resources Research* 12: 127–134. https://doi.org/10.1023/A:1024214812527

Pleasants, J. M. 2015. "Monarch Butterflies and Agriculture." In *Monarchs in a Changing World: Biology and Conservation of an Iconic Butterfly*, edited by K. S. Oberhauser, K. R. Nail, and S. M.Altizer. Ithaca, NY: Cornell University Press.

United States Department of Agriculture. 2019. "Adoption of Genetically Engineered Crops in the U.S.: Recent Trends in GE Adoption." USDA Economic Research Service, Data Products. *https://www.ers.usda.gov/data-products/adoption-of-genetically-engineered-crops-in-the-us/recent-trends-in-ge-adoption*

SPRING TO SUMMER

Prudic, K. L., S. Khera, A. Sólyom, and B. N. Timmermann. 2007. "Isolation, Identification, and Quantification of Potential Defensive Compounds in the Viceroy Butterfly and its Larval Host–Plant, Carolina Willow." *Journal of Chemical Ecology* 33 (6): 1149–59. https://doi.org/10.1007/s10886-007-9282-5

Ritland, D. B., and L. P. Brower. 1991. "The Viceroy Butterfly is not a Batesian Mimic." *Nature* 350 (6318): 497–498. https://ui.adsabs.harvard.edu/abs/1991Natur.350..497R/abstract

United States Department of Agriculture. https://www.fsa.usda.gov/programs-and-services/conservation-programs/conservation-reserve-program

A SUMMER BREAK TO BIKE

De Anda, A., and K. S. Oberhauser. 2015. "Invertebrate Natural Enemies and Stage-Specific Mortality Rates of Monarch Eggs and Larvae." In *Monarchs in a Changing World: Biology and Conservation of an Iconic Butterfly*, edited by K. S. Oberhauser, K. R. Nail, and S. M. Altizer. Ithaca, NY: Cornell University Press.

Monarch Joint Venture. https://monarchjointventure.org/resources/faq/natural-enemies

THE NORTH LANDS

Monarch Watch. https://monarchwatch.org/blog/2020/06/03/monarch-annual-cycle-migrations-and-the-number-of-generations/

Zhu, H., R. J. Gegear, A. Casselman, S. Kanginakudru, and S. M. Reppert. 2009. "Defining Behavioral and Molecular Differences Between Summer and Migratory Monarch Butterflies." *BMC Biology* 7 (14). https://doi.org/10.1186/1741-7007-7-14

WELCOMING "WEEDS"

Agrawal, A. A., J. G. Ali, S. Rassman, and M. Fishbein. 2015. "Macroevolutionary Trends in the Defense of Milkweeds Against Monarchs." In *Monarchs in a Changing World: Biology and Conservation of an Iconic Butterfly*, edited by K. S. Oberhauser, K. R. Nail, and S. M. Altizer. Ithaca, NY: Cornell University Press.

Koh, I., E. V. Lonsdorfa, N. M. Williams, C. Brittain, R. Isaacs, J. Gibbs, and T. Ricketts. 2015. "Modeling the Status, Trends, and Impacts of Wild Bee Abundance in the United States." *Proceedings of the National Academy of Sciences* 113 (1): 140–145. https://doi.org/10.1073/pnas.1517685113

Wozniacka, G. 2013. "Beekeepers, Environmentalists Sue EPA for Not Suspending Pesticides That May Harm Bees." *Associated Press*, March 21. http://www.columbia.org/pdf_files/centerforfoodsafety184.pdf

ALONG THE ATLANTIC

Bartel, R. A., K. S. Oberhauser, J. C. de Roode, and S. M. Altizer. 2011. "Monarch Butterfly Migration and Parasite Transmission in Eastern North America." *Ecology* 92 (2): 342–351. https://doi.org/10.1890/10-0489.1

Faldyn, M. J., M. D. Hunter, and B. D. Elderd. 2018. "Climate Change and an Invasive, Tropical Milkweed: an Ecological Trap for Monarch Butterflies." *Ecology* 99 (5): 1031–1038. https://esajournals.onlinelibrary.wiley.com/doi/full/10.1002/ecy.2198

Majewska, A. A., and S. Altizer. 2019. "Exposure to Non-Native Tropical Milkweed Promotes Reproductive Development in Migratory Monarch Butterflies." *Insects* 10 (8): 253. https://doi.org/10.3390/insects10080253

Monarch Joint Venture. https://monarchjointventure.org/images/uploads/documents/OE_fact_sheet_Updated.pdf

O. R. Taylor, Jr. Personal communication with author, June 17, 2020.

Reppert, S. M., and J. C. de Roode. 2018. "Demystifying Monarch Butterfly Migration." *Current Biology* 23: 1009–1022. https://doi.org/10.1016/j.cub.2018.02.067

Satterfield, D. A., J. C. Maerz, and S. Altizer. 2015. "Loss of Migratory Behaviour Increases Infection Risk for a Butterfly Host." *Proceedings of the Royal Society B* 282 (1801). http://doi.org/10.1098/rspb.2014.1734

Stager, J. C., B. Wiltse, J. B. Hubeny, E. Yankowsky, D. Nardelli, and R. Primack. 2018. "Climate Variability and Cultural Eutrophication at Walden Pond (Massachusetts, USA) During the Last 1800 Years." *PLOS ONE* 13 (4): e0191755. https://doi.org/10.1371/journal.pone.0191755

Taylor, O. R. Jr., J. P. Lovett, D. L. Gibo, E. L. Weiser, W. E. Thogmartin, D. J. Semmens, J. E. Diffendorfer, J. M. Pleasants, S. D. Pecoraro, and R. Grundel. 2019. "Is the Timing, Pace, and Success of the Monarch Migration Associated with Sun Angle?" *Frontiers in Ecology and Evolution* 7: 442. https://doi.org/10.3389/fevo.2019.00442

BACK TOWARD CANADA

Hristov, N. I., and W. E. Conner. 2005. "Sound Strategy: Acoustic Aposematism in the Bat–Tiger Moth Arms Race." *Naturwissenschaften* 92 (4): 164–169. https://doi.org/10.1007/s00114-005-0611-7

CANADA, TAKE TWO

Darlene Burgess. Personal communication with author, June 15, 2020.

Don Davis. Personal communication with author, June 15, 2020.

Flockhart, D. T. T., B. Fitz-gerald, L. P. Brower, R. Derbyshire, S. Altizer, K. A. Hobson, L. I. Wassenaar, and D. R. Norris. 2017. "Migration Distance as a Selective Episode for Wing Morphology in a Migratory Insect." *Movement Ecology* 5 (7). https://doi.org/10.1186/s40462-017-0098-9

Freedman, M., and H. Dingle. 2018. "Wing Morphology in Migratory North American Monarchs: Characterizing Sources of Variation and Understanding Changes Through Time." *Animal Migration* 5 (1): 61–73. http://doi:10.1515/ami-2018-0003

Journey North. https://journeynorth.org/tm/monarch/DavisDonBio.html

Journey North. https://journeynorth.org/tm/monarch/DiscoveryTale.html

Micah Freedman. Personal communication with author, January 8, 2020.

Monarch Watch. https://monarchwatch.org/press/press-briefing.html

Schroeder, H., A. Majewska, and S. Altizer. 2020. "Monarch Butterflies Reared Under Autumn-Like Conditions Have More Efficient Flight and Lower Post-Flight Metabolism." *Ecological Entomology 45 (3): 562–572* https://doi.org/10.1111/een.12828

Urquhart, F. 1976. "Found at Last: the Monarch's Winter Home." *National Geographic* 150 (2): 160–174.

MAINTAINING THE LEAD

Sacchi, C. 1987. "Variability in Dispersal Ability of Common Milkweed, *Asclepias syriaca*, Seeds." *Oikos* 49 (2): 191–198. http://doi:10.2307/3566026

THE OUT-OF-THE-WAY WAY

Sáenz-Romero, C., G. E. Rehfeldt, P. Duval, and R. A. Lindig-Cisneros. 2012. "Abies Religiosa Habitat Prediction in Climatic Change Scenarios and Implications for Monarch Butterfly Conservation in Mexico." *Forest Ecology and Management* 275: 98–106. https://doi.org/10.1016/j.foreco.2012.03.004

Waldman, S. 2018. "2017 Was the Third Hottest Year on Record for the U.S." *Scientific American*: January 9. https://www.scientificamerican.com/article/2017-was-the-third-hottest-year-on-record-for-the-u-s/

STRAGGLING WITH STRAGGLERS

Alonso-Mejia, A., E. Rendón-Salinas, E. Montesinos-Patino, and L. Brower. 1997. "Use of Lipid Reserves by Monarch Butterflies Overwintering in Mexico: Implications for Conservation." *Ecological Applications* 7 (3): 934–947. https://doi.org/10.2307/2269444

Masters, A. R., S. B. Malcolm, and L. P. Brower. 1988. "Monarch Butterfly (*Danaus plexippus*) Thermoregulatory Behavior and Adaptations for Overwintering in México." *Ecology* 69 (2): 458–467. https://doi.org/10.2307/1940444

O. R. Taylor, Jr. Personal communication with author, June 14, 2020.

Taylor, O. R. Jr., J. P. Lovett, D. L. Gibo, E. L. Weiser, W. E. Thogmartin, D. J. Semmens, J. E. Diffendorfer, J. M. Pleasants, S. D. Pecoraro, and R. Grundel. 2019. "Is the Timing, Pace, and Success of the Monarch Migration Associated with Sun Angle?" *Frontiers in Ecology and Evolution* 7: 442. https://doi.org/10.3389/fevo.2019.00442

THE SOUTHERN REWIND

Guerra, P. A., and S. M. Reppert. 2013. "Coldness Triggers Northward Flight in Remigrant Monarch Butterflies." *Current Biology* 23: 419–423. https://doi.org/10.1016/j.cub.2013.01.052

Reppert, S. M., and J. C. de Roode. 2018. "Demystifying Monarch Butterfly Migration." *Current Biology* 23: 1009–1022. https://doi.org/10.1016/j.cub.2018.02.067

Steven Reppert. Personal communication with author, December 12, 2019.

CROSSING THE BORDER

J. L. Tracy. Personal communication with author, June 21, 2020.

Kantola, T., J. L. Tracy, K. A. Baum, M. A. Quinn, and R. N. Coulson. 2019. "Spatial Risk Assessment of Eastern Monarch Butterfly Road Mortality During Autumn Migration Within the Southern Corridor." *Biology Conservation* 231: 150–160. https://doi.org/10.1016/j.biocon.2019.01.008

Mora Alvarez, B. X., R. Carrera-Treviño, and K. A. Hobson. 2019. "Mortality of Monarch Butterflies (*Danaus Plexippus*) at Two Highway Crossing "Hotspots" During Autumn Migration in Northeast Mexico." *Frontiers in Ecology and Evolution* 7: 273. https://doi.org/10.3389/fevo.2019.00273

Acknowledgments

My bike tour, solo in design, was a giant group effort. Alone, I would have passed all my nights in my tent, showered disgustingly fewer times, and had exponentially less ice cream. Most important, my voice on behalf of the monarchs would have been a mere whisper. There are more people to thank than there are miles in this story.

Thanks to everyone who invited me into their homes and gifted me meals, showers, beds, internet, towels (oh, the luxury of a fluffy towel), and every detail that made each visit its own. Such invitations were tokens of trust and metaphorical pats on the back. I wrote about each and every one of you in the first draft of this book. Then, acknowledging the word count, I began the painful process of whittling down. Know that while I couldn't include every account of hospitality in my story, you are all in its DNA. It was your collective support, inspiration, friendship, beauty, generosity, revolutionary spirt, and nurturing example that allowed every mile and informed every sentence. Thanks to all of you: Brianda Cruz González, Leticia González Valencia, *y la familia*; Moises Acosta and Veronica Velazquez Araujo; Vicky; Francisco Martínez García, *y la familia*; Rodolfo *y la familia*; David Forbes PhD; Emily, Jan and Bill Neiman; Elizabeth and Stan Hart; Linda Lavender, Mike Cochran, Meg and Eva; Sandy and Mike Schwinn; Amy, Mike, Drew, and Hannah Whitaker; Jenny Reynolds; Patrick, Alice, and Leo Martin; Patty and Gary Dykman; Angie, Katie, Allison, and Kevin Babbit; Kate, Emma, Megan, and Annie Rezac; Susan Allen; Kim and Tony Nunez; Jeanne; Tom, Sarah, and Clyde Ehrhardt; Andrew and Phyllis Schmid; Sue and Chris Eidem; Maddy Cochran; Val Biron; Dave Lickley and Heather Jeramaz; Dr. Jane Cox and Dr. Gary Bota; Margaret and Brian; Candy and Peter Campbell; Nancy and John Hayden; Jess Huyghebaert; Katie, Brian, Lizzie, and Poppy Hone; the field technicians at the Assabet River NWR field house; Dan Fagin and Alison Frankel; Pat Wright;

Erica Lim, Obe Ben Samuel, and Lihn Nguyen; Mitch; Don Bradford and Bruce England; Heather, Dan, Mia, Lucie, Jude, and J. P. Schieder; Barbara Deharte; Thanh-Thanh Tieu and Mark Mollison; Barb and Mark Hacking; Bruce Parker and Jinny Behrens; Darlene and Ken Burgess; Louie Fiorino; Louise Weber; Kylee and Romie Baumle; Dana and family; Gena Garrett; Bonnie, Marc, and Jacob Mahnke; Rick and Connie Mihalevich; Candy and family; Stephanie Michels and Everett Stokes; Nadia Navarrete-Tindall and Randy Tindall; Bill Fessler; Marjie and Steve Dykman; Delia Lister; Val and Stash Frankoski; Kate Barnes; Tom and Cindy Rimkus; Nathan Nunn; Felicity and Paul Gatchell; Annie the Cat; Elizabeth, Brandon, Emma, Charlie, and Teddy McBride; Allison Jackson; the *tamal*-making family; Iván Cumplián Medlin; Elda Margarita Villasana Rojas, Carlos Adrián Verdin Licon, Carlos and Diego; and Chloe. A double thanks to the Neimans and the Martins, with whom I was lucky enough to stay in both the spring and the fall.

A quadruple gracias to not only Brianda, but to everyone in Michoacán who gave me a home for four winters and counting. To Brianda, your patience and acceptance is heroic, your invitation, a blessing. Thank you, Leticia, for being a wonderful example of strength and love. To the rest of the family—Israel, Ivan, Diana, Israelito, Mari, Fer (and Dobber)—for sharing your home. Thanks to Gloria, Pablo, Lola, Edwin, Juanito, Fabian, Vanesa, Richard, Leo, Jorge, Fabi, Roberto, Paulino, Graciela, Maura, Paulina, Estela, and your families for inviting me in as well. And to the guides at El Rosario and Papalotzin, for letting me loiter, bend rules, and become an honorary local.

And countless thanks to my Kansas City hosts (and parents!), Marjie and Steve Dykman. You spoil me on every visit. I'm lucky to have such a rest stop. Thanks for cheering me on, dealing with my crazy choices, and not worrying *too* much. If you think about it, it is actually *your* fault that this adventure came to be. You, after all, let me raise frogs in my bedroom, brought Gatorade to all my cross-country meets, taught me how to ride a bike, signed me up for art classes, embraced my college of choice, and (most influentially) made my brother and me swim across that (potentially) alligator-infested pond.

While invitations to shower were coveted, opportunities to speak were victories. Thanks to the students who attended my presentations, for giving me your attention, brilliant questions, and pure amazement (and for agreeing that American toads are the cutest of creatures). To the teachers, for gifting me precious hours of classroom time. To everyone who put me in touch, spread the word, and made each of the following school visits happen (especially my main contacts): Southwest Texas Junior College, Del Rio, TX (David Forbes); Pegasus School of Liberal Arts, Dallas, TX (Elizabeth Hart); Newton Rayzor Elementary, Denton, TX (Linda Lavender); Leisure Park Elementary, Broken Arrow, OK (Sandy Schwinn and Sheila Reid Schulz); Town and Country, Tulsa, OK (Amy Lucas Whitaker); schools in Wichita, KS (Lori Jones); Pleasanton Elementary, Pleasanton, KS; LaCygne Elementary, LaCygne, KS; Rockville Elementary and Broadmoor Elementary, Louisburg, KS (Patrick Martin); Tomahawk Elementary, Overland Park, KS (Brian Watson); Timber Creek and Indian Creek Elementary, Overland Park, KS (Christine Gold); Frank Ruston Elementary, Kansas City, KS (Peter Wetzel); Merriam Park, Merriam, KS (Heidi Walker); Hale Cook, Kansas City, MO (David Darmitzel); Citizens of the World Charter School, Kansas City, MO (Andrew Johnson); St. Margaret Mary, Omaha, NE (Kate Rezac); West Harrison Elementary, Mondamin, IA (Kim Nunez); Spalding Park Elementary, Sioux City, IA (Mande Moran and Mimi Moore); Algonquin Road Public School, Sudbury, ON (Petra Demeyere); North Berwick Elementary, North Berwick, ME (William Fulford); Mighty Oak Elementary, Lakeshore, ON (Lindsay Logsdon); Wayne Trace Payne Elementary, Payne, OH (Kylee Baumle); Harper Elementary, Holy Spirit School, and Vogel Elementary, Evansville, IN (Gena Garrett); St. Vincent de Paul, Cape Girardeau, MO (Bonnie Mahnke); Mill Creek Elementary, Columbia, MO (Megan Kinkade); Academie Lafayette, Kansas City , MO (Krista Story and Céline Ghisalberti); Louisburg Middle, Louisburg, KS (Mike Isaacsen and Patrick Martin); Cecil Floyd Elementary, Thomas Jefferson Independent, Royal Heights Elementary, and St. Mary's Elementary, Joplin, MO (Val Frankoski); Central Elementary, Neosho, KS (Bruce Hallman); Owl Creek Middle, Fayetteville, AR (Matt Pledger and Kate Barnes); Huntsville

Intermediate, Huntsville, AR (Sarah Glenn); Regents Austin, Austin, TX (Allison Jackson); Palm Elementary, Austin, TX (Mario Vasquez); Menchaca Elementary, Austin, TX (Lucretia Beard); Consuelo Mendez Middle, Austin, TX (Sherry Lepine); Joslin Elementary, Austin, TX (Kate Mason-Murphy and Summer McKinnon); Brentwood, Austin, TX (Theresa Wood); Eastside Memorial Early College, Austin, TX (Rhonda Barton); Kealing Middle and Highland Park Elementary, Austin, TX (Elizabeth McBride); Primaria Benito Juárez, Jaumave, Tamps (Benjamín Hernandez).

Thanks to Patrick Martin, manager of Marais Des Cynges National Wildlife Refuge for pushing me to think bigger. You reached out early in the planning process with an idea to connect me to the network of Wildlife Refuges, so that I might organize public presentations. I loved the idea, and with your help (and later a web of contacts across the country, including Melissa Clark and Becky Lungenecker), more nature centers (and other venues) reached out with interest.

Thanks to every venue that opened its doors, the staff and volunteers who planned and promoted each event, and all the folks that came out, offered donations, laughed at my corny jokes, and passed along what they learned: Hagerman NWR, Sherman, TX (Courtney Anderson); Great Plains Nature Center, Wichita, KS (Lori Jones); Monarch Watch, Lawrence, KS (Chip Taylor and Angie Babbit); Loess Bluffs NWR, Mound City, MO (Lindsey Landowski); Swanson Branch Library, Omaha, NE (Nancy Chmiel); DeSoto and Boyer Chute National Wildlife Refuges (Peter Rea); Northwest Iowa Group Sierra Club, Sioux City (Jeanne); Prairie Heritage Center, Peterson, IA (Charlene Elyea and Becca Castle); Water's Edge Nature Center, Angola, IA (Julie Fosado); Minnesota Valley National Wildlife Refuge and Lake Nokomis Community Center, Minneapolis, MN (Samantha Herrick); Living with Lakes Centre, Sudbury, ON (John Gunn); Varnum Library, Jeffersonville, VT (John and Nancy Hayden); Fairbanks Museum, St. Johnsbury, VT (Steve Agius and Leila Nordmann); Ipswich Open Space Committee and Ispwich Town Hall, Ispwich, MA (Katie Banks Hone); Assabet River NWR, Sudbury, MA (Jared Green); Royal Botanical Gardens, Burlington, ON (Karin Davidson-Taylor); Greenway Garden Centre, Breslau, ON

(Thanh-Thanh Tieu); Communities in Bloom and the Local Community Food Center, Stratford, ON (Barb Hacking); Point Peelee National Park Visitor Center, Leamington, Ontario (Andrew Laforet); University of Saint Francis and Little River Wetlands Project, Fort Wayne, IN (Renee Wright and Kylee Baumle); Cope Environmental Center, Centerville, IN (Aubrey Blue); Sierra Club's Southwest Indiana Network and the Evansville chapter of Navigators USA, Evansville, IN (Gena Garrett); Woodlands Nature Station, Land Between the Lakes National Recreation Area, Cadiz, KY (John Pollpeter); Mizzou Botanic Garden, Columbia, MO (Karlan Seville, Megan Tyminski, and Caroline Dohack); Big Muddy Speaker Series and Les Bourgeois Vineyards Bistro, Rocheport, MO (Steve Schnarr and Tim Haller); Big Muddy Speaker Series and Anita B. Gorman Conservation Discovery Center, Kansas City, MO (Larry O'Donnell and Michael Morgan); Nature Reach, Pittsburg State University, Pittsburg, KS (Delia Lister); Wildcat Glades Conservation & Audubon Center, Joplin, MO (Val Frankoski); and Lady Bird Johnson Wildflower Center, Austin, TX (Tanya Zastrow).

Every presentation came to fruition because of a chain of people, often invisible to me. Such connections meant that I received more offers than I could accept. Thanks to everyone who got in touch, and an even bigger thanks for understanding when I couldn't make a visit happen.

My thanks now extend from *riding* the miles to *writing* about the miles.

I wrote much of this book at Brianda's house, Papalotzin, my parents' house, and in Tom and Debra Wiestar's cabin, among the cedars and pines in the foothills of the Sierra. Thanks, Tom and Deb, for the refuge, the guidance, and the unyielding support. Thanks for the elaborate lighting for my desk, the constant supply of delicious food and poison oak remedies, and for partaking in my marathon conversations, when, after days alone, I needed people to talk to. I am lucky to have such role models.

As I wrote, my friends became editors. Thanks to Jenny Long for reading the whole dang thing, even though you don't like bugs. To Debra Weistar, Nia Thomas, Chrissa Pedersen, Chip Taylor, and Davin Hart for reading all of it, part of it, or excerpts that became magazine articles. Thanks to all my hosts that I asked to be editors and to help make sure I got the facts of my stay correct. Thanks to Stacee Lawrence and Julie Talbot, editors at Timber Press, for smoothing out the edges of my rough draft. Extra thanks to Julie for being more than an editor. Your patience, guidance, and support were the writing equivalent of biking with a tailwind. Thanks to the editors (Kirsten Traynor and Bryan Dykman), writers (Dan Faggin, Katie Yale, and Kylee Baumle), and publishers (Carol Malnur), for your advice. To Kira Miller and the SNAMP frog team for letting me process the many stages of this book as we tromped through the woods.

While I wrote, I also leaned on monarch scientists. Broadly speaking, I want to thank every monarch scientist—heck, every scientist—for pointing us to truths, guiding our understanding, and holding humanity accountable for our actions. You are heroes, and so it was always a wonderful shock and honor to receive replies to my many questions and feedback on my many drafts. Thanks to Steve Reppert, Cuauhtémoc Sáenz-Romero, David Gibo, Micah Freedman, Eligio García Serrano, James Tracy, Mónica Missrie, David Bray, Don Reynolds, Catherine Tucker, and especially to Chip Taylor. Chip, your continued support and dedicated example are pillars of this project. It has been an immense privilege to learn from you.

Thanks to my fellow past-trip adventurers—Galen Reid, Tommy Viducich, Aaron Viducich, Matt Shift, Alyssum Cochen, Nia Thomas, and Matt Titre—for pushing my boundaries. To my fellow wildlife technician adventurers for encouraging our outrageous devotion. To my fellow renegades at GreenWheels, the Campus Center for Appropriate Technology, and Synergia for showing me how to better our world.

Thanks to all my teachers. The education you provided was a privilege linked to all my successes. It was an honor to see some of you at my presentations and to present in some of your classrooms. Thanks to Mr. Lockard for showing me that loving animals was not so impossible and to Mr. Johnson for being a brilliant example of a scientist with an opinion.

Then there are the strangers connected by invisible threads that deserve my thanks. To all the drivers who slowed down and passed me with caution. To NPR and podcast hosts whose familiar voices kept me company on some long, lonely stretches. To Terry Tempest Williams, Robin

Wall Kimmerer, Sy Montgomery, and the many other authors whose words are comfort and make the world a better place.

Thanks to everyone fighting, in endlessly big and small ways, on behalf of the monarchs and their neighbors. Our paths may not have crossed, but your efforts are seen, felt, and appreciated. Biking past an unmowed ditch or a lawn devoted to natives will always make me hoot with joy.

And finally, with all my heart and soul, thanks to the monarchs. You amaze me. You have become my teachers, encouraging an adventure, teaching me Spanish, watercolor, web design, video editing, photography, networking, public speaking, gardening, stewardship, science, and love. You helped me write this book, and every word of it is for you.

Index